Topical Conference: Decarbonization of the Chemical Industry through Electrification

Held at the 2023 AIChE Annual Meeting

Orlando, Florida, USA
5-10 November 2023

ISBN: 978-1-7138-9298-4

Printed from e-media with permission by:

Curran Associates, Inc.
57 Morehouse Lane
Red Hook, NY 12571

Some format issues inherent in the e-media version may also appear in this print version.

Printed with permission by Curran Associates, Inc. (2024)

For permission requests, please contact AIChE
at the address below.

AIChE
120 Wall Street, FL 23
New York, NY 10005-4020

Phone: (800) 242-4363
Fax: (203) 775-5177

www.aiche.org

Additional copies of this publication are available from:

Curran Associates, Inc.
57 Morehouse Lane
Red Hook, NY 12571 USA
Phone: 845-758-0400
Fax: 845-758-2633
Email: curran@proceedings.com
Web: www.proceedings.com

TABLE OF CONTENTS

DECARBONIZATION BY ELECTRIFICATION: REACTORS AND REACTIONS 1

DECARBONIZATION BY ELECTRIFICATION: SEPARATIONS

DECARBONIZATION BY ELECTRIFICATION: SYSTEMS AND POLICY

DECARBONIZATION BY ELECTRIFICATION: REACTORS AND REACTIONS 2

DECARBONIZATION BY ELECTRIFICATION: REACTORS AND REACTIONS 3

DECARBONIZATION BY ELECTRIFICATION

POSTER SESSION: DECARBONIZATION BY ELECTRIFICATION

222a Electrification of Steam Methane Reforming By Induction Heating

Chengtian Cui, Yufei Zhao, Cornelius Masuku

The decarbonization of the chemical industry is of significance to mitigate global warming. The use of carbon-free electrification is a way of decarbonizing the chemical industry. In this work, modeling design of an electrified steam methane reformer (SMR) is investigated. The SMR features strong endothermic reactions, and conventionally driven by the combustion of natural gas. Herein, induction heating generated by an alternating magnetic field is applied to provide a non-contact heat source to the catalyst sites. Among different induction heating mechanisms, eddy current and hysteresis were considered for heating purpose. A first principle mathematical model was developed by considering the fluid mass, energy, momentum balances as well as diffusion-reaction, both inside (catalyst domain) and outside (bulk gas phase) of the catalyst pellet. Different scale of models was built by considering one-dimensional or two-dimensional model, accounting for diffusion or dispersion, and heterogeneous or pseudo-homogeneous reactions, among others. The results indicate that the electrified reformer is more compact, energy-efficient, and has faster dynamics than the current reformers.

222b Non-Precious Electrodes Design for Hydroxide Exchange Membrane Electrolyzers

Fei Xiao, Alexandra Oliveira, Brian P. Setzler, Minhua Shao, Yushan Yan

Hydrogen is an excellent energy carrier to replace fossil fuels with renewable energy and the green hydrogen generated by electrolysis is the cleanest candidate.[1-2] Department of Energy in the United States has set '1 dollar ($) per kilogram (kg) in one-decade' cost and 80-thousand-hour durability targets for clean hydrogen production. The current price of green hydrogen (5-6 $/kg) must be lowered by reducing electricity and electrocatalyst expense. The increased capacity of renewable energy leads to a significant price reduction in electricity.[3] The next task is to cheapen the material cost of catalysts. Because various non-precious metal materials and components are stable in high pH environments, hydroxide exchange membrane electrolyzers (HEMELs) can use non-precious metal catalysts in both electrodes. However, the weakly adhered catalyst layers in HEMELs always suffer from continuous gas generation and bubble detachment disturbance, resulting in a short lifespan.

Herein, we report a work about a novel cathodic electrode preparation technique to improve the activity and durability of non-precious metal based HEMELs. Propylene glycol is added into the conventional water and alcohol solvent system to enhance the adherence among Ni-Mo/Engineered catalyst support catalyst (purchased from Pajarito Powder company) agglomerates. An additional carbon layer configuration is designed between the gas diffusion and catalyst layers to alleviate wash-off and transfer-to-membrane issues during tests. This HEMEL with Ni-Fe-OOH anode[4] and Ni-Mo/Engineered catalyst support cathode can achieve a 500 mA cm^{-2} current density at 1.8 V under pure water. The optimized electrode structure can prevent the detachment of catalysts during continuous operation and demonstrate good voltage stability at 200 mA cm^{-2} current density for above 100 h. The role of propylene glycol in the catalyst-ink formulation and the contribution of the additional carbon layer to water/gas transport efficiency will be systematically explored. This work highlights the importance of electrode design for achieving highly active and stable non-precious metal based HEMELs, and sheds light on developing advanced electrode structures for other electrochemical devices.

1. Oliveira, A. M.; Beswick, R. R.; Yan, Y., *Current Opinion in Chemical Engineering* **2021,** *33*, 100701.
2. Abbasi, R.; Setzler, B. P.; Lin, S.; Wang, J.; Zhao, Y.; Xu, H.; Pivovar, B.; Tian, B.; Chen, X.; Wu, G.; Yan, Y., *Advanced Materials* **2019,** *31* (31), 1805876.
3. Pivovar, B.; Rustagi, N.; Satyapal, S., *The Electrochemical Society Interface* **2018,** *27* (1), 47.
4. Xiao, J.; Oliveira, A. M.; Wang, L.; Zhao, Y.; Wang, T.; Wang, J.; Setzler, B. P.; Yan, Y., *ACS Catalysis* **2021,** *11* (1), 264-270.

222c Design Rules for Generating Acid and Base from Membrane-Free Electrolysis of Brine

Daniela Fraga Alvarez, Dimitri Livitz, Xueqi Pang, Nafis Mahmud, Daniel Esposito, Kyle Bishop, Muftah El-Naas

As water scarcity issues become exacerbated by climate change and population growth, an evident solution is to enlarge potable water production via desalinization. Unfortunately, this technology leads to major environmental side effects that result from the discharge of concentrated seawater (reject brine) into neighboring bodies of seawater, land, or evaporation ponds. The long-term impact of this practice on marine life is still unknown, but the brine's high temperature and salinity levels could be devastating. On the other hand, reject brine contains minerals and salts like calcium, magnesium, and sodium chloride that can be harvested and reutilized by other industries. A promising approach to treat reject brine is to use membraneless electrolyzers, which are well-suited for operation with highly conductive liquid electrolytes and avoid the need for membranes by controlling fluid flow to separate anode and cathode products before they can cross over to opposing effluent streams. They can operate in a wide range of pH environments and simultaneously generate acid and base; however, the lack of separator can result in high levels of product crossover if the electrolyzer is not designed or operated properly. This issue can be addressed by operating at higher flow rates, which can push bubbles away from the centerline separating the two electrodes, thereby minimizing crossover. However, higher flow rates also dilute the generated product. In this study, we investigate the trade-offs between crossover, cell voltage, and the concentration of generated acid and base with a combination of electroanalytical measurements, *in situ* high-speed videography, pH dye experiments, and modeling. Additionally, the distribution of catalysts was varied to control where the product was generated and further reduce crossover. Our study demonstrates that placing the active catalyst in the opposite face of the electrodes increases current utilization by up to 40% and improves catalyst stability. Further work involves the scale-up and optimization of the reactor design.

222d System Design and Testing of Bzcy Proton Conducting Ceramic Membranes for Electrochemical Steam Methane Reforming at High Temperature and Pressure

Derek Richard, Berkay Citmaci, Junwei Luo, Joonbaek Jang, Panagiotis Christofides, Carlos Morales-Guio

Hydrogen is traditionally produced through steam methane reforming (SMR) where natural gas and steam are reacted at high temperatures to produce a mixture of CO_2, CO, and H_2 from which H_2 must be separated. Due to the multiple reaction and separation steps involved and the need for heat recovery integration, SMR must be conducted at large scales in centralized plants to reach profitable efficiencies. Alternatively, water electrolyzers can be used to electrochemically produce H_2 from water. However, the minimum energy required to produce H_2 from electrolysis is still almost seven times higher than from SMR. A promising alternative process is the use of proton conducting ceramic $BaZr_{0.8-x-y}Ce_xY_yO_{3-\delta}$ (BZCY) membranes to perform SMR electrochemically. Combining the SMR process with these proton conductive ceramics enables H_2 separation and compression to be integrated into the reaction process through electrochemical pumping of the protons. The removal of H_2 also provides the benefit of shifting the equilibrium of the SMR reaction so that nearly complete conversion of methane can be reached. Additionally, the separation and compression of H_2 generates heat directly in the membrane that can be balanced with the energy consumed by the endothermic SMR reaction to provide micro-scale heat integration [1]. This technology has the potential to transform how H_2 is generated, and by pairing other reactions with the separated protons, alternative and more direct pathways to produce valuable chemicals such as ammonia can be developed [2].

In this study, we present the details of a system developed in our lab to test the performance of these proton conducting ceramic membranes and present the results of our study under a wide range of pressure, temperature, and gas compositions. Detailed considerations of the system design include pressures up to 50 bar, temperatures up to 900°C, and electrical connections to the membrane. The details of how sensing and control were designed to allow for automated data collection are also discussed [3]. Additionally, results from tests performed on membrane assemblies provided by CoorsTek Membrane Sciences are presented for proton pumping and SMR. Lastly, the data from these tests is used to develop a reduced order model that can be used to inform engineering decisions in larger scale system design.

References:

1. Malerød-Fjeld, H., Clark, D., Yuste-Tirados, I., Zanón, R., Catalán-Martinez, D., Beeaff, D., Morejudo, S. H., Vestre, P. K., Norby, T., Haugsrud, R., Serra, J. M., Kjølseth, C., 'Thermo-electrochemical production of compressed hydrogen from methane with near-zero energy loss,' *Nature Energy* 12 (2017): 923-931.
2. Kyriakou, V., Garagounis, I., Vourros, A., Vasileiou, W., Stoukides, M., 'An electrochemical Haber-Bosch process,' *Joule* 1 (2020): 142-158.
3. Çıtmacı, B., Luo, J., Jang, J., Korambath, P., Morales-Guio, C., Davis, J., Christofides, P. D., 'Digitalization of an experimental electrochemical reactor via the smart manufacturing innovation platform,' *Digital Chemical Engineering* 5 (2022): 100050.

222e Cold Plasma Enhanced Dry Methane Reforming Reactor and Process Design to Maximize the Overall Process Sustainability

Srinivas Rangarajan, Diego Gonzalez, Tian Qin, Jonas Baltrusaitis, Wei-Min Huang

Room temperature plasma is an attractive method to produce syngas via electrification of the endothermal methane dry reforming reactors. Electricity can be obtained from green sources thus lowering the CO_2 burden of the reactor that normally utilizes fixed heat flux generates by burning natural gas. The optimization of such reactor operating conditions, however, becomes more complex and laboratory experiments need to be performed in conjunction with the experiment optimization to account for the multitude of external variables (chiefly thermal heat vs electricity to generate plasma), the resulting performance parameters, such as generated H_2 amount, as well as sustainability constrains, such as the source of electricity used.

In this presentation, we show a comprehensive plasma-enhanced catalytic dry methane reforming reactor analysis of experiments to obtain the optimal operating conditions with the objective to maximize H_2 production while minimizing the associated environmental impacts. In particular, variables explored included CO_2/CH_4 ratio, gas flow, reactor temperature, plasma intensity and water vapor content. The response parameters included reactant conversion, H_2 produced, rate of reaction and carbon balance. Importantly, environmental impact calculations were also performed using Life Cycle Analysis to calculate environmental burden, in kg CO_2 equivalent using the laboratory data obtained during the experiments. These calculations resulted in a set of solutions that allowed to maximize H_2 production while minimizing their environmental impacts. Practical process design parameters that utilized these data will also be discussed.

222f Electrified Steam Methane Reforming with High Frequency Magnetic Induction

Jonathan Fan, Calvin Lin, Pinak Mohapatra, Zhennan Ru, Ariana Hofelmann, Dolly Mantle, Chenghao Wan

The chemical industry has been one of the major contributors of greenhouse gas (GHG) emissions, accounting for nearly one third of global CO_2 emissions. Many of these industrial processes require the heating of volumetric reaction media to high temperatures, with applications ranging from the selective capture of gases (e.g., CO_2, SO_2, etc) and the production of fuels and chemicals (e.g., biomass decomposition, hydrogen and syngas generation, etc) to the processing of materials (e.g., baking of bricks, sintering of minerals, etc). One solution to mitigating these carbon emissions is to utilize renewable and clean electricity, instead of fossil fuels, to provide endothermic heat for these processes. Amongst the wide range of environmentally friendly electrification heating concepts currently being explored, inductive heating is an established method for clean, efficient, and rapid heating and is used industrially in the processing of raw iron and silicon materials. However, an underlying challenge to date is understanding how inductive heating can be adapted to heat volumetric media uniformly and its integration with different reactor configurations. An extensive and thorough understanding of the design of three-dimensional heating susceptors, catalyst integration with the susceptor, and co-design of power electronics with the susceptor are required.

In this study, we discuss the potential for inductive heating to serve as an electrification heating solution for heating steam methane reforming catalytic reactors. Metallic open cell lattice susceptors were washcoated with the zirconia and impregnated with Ni solution to act as both heating susceptors and supported catalyst material. With a combination of analytic and numerical electromagnetic optimization, the lattice geometry was optimized to produce uniform volumetric heating across the susceptor, minimizing temperature gradients within the reactor and minimizing heat transfer limitations between the susceptor and catalyst. A focal point of our research effort was in addressing heating efficiency, which has been limited in prior studies by poor energy transfer efficiency between the magnetic coils and susceptors due to operation at low frequencies. In combination with parametric studies on the effects of circuit frequency and coil geometry on heating efficiency, we utilize a high frequency inductive heating scheme, based on megahertz frequency power electronics, to enable susceptor heating at near unity efficiencies. We evaluate the reactor system's performance through experiments at different residence time and temperature and experimentally demonstrate that operating the setup at MHz scale frequency can lead to heating efficiencies above 90%, comparable to resistive heating method.

222g Feasibility Study on the Resistive Heating of Catalytic Fecral Wire for Steam Methane Reforming

Elmer Ledesma, Meghana Idamakanti, Praveen Bollini, Vemuri Balakotaiah, Michael Harold

The global hydrogen generation market, valued at USD 148 billion in 2021, is forecasted to reach USD 280 billion by 2030. This growth and demand for hydrogen is driven principally by ammonia and methanol production, and oil refining. The largest share of this market revenue comes from steam methane reforming (SMR). Compared to other commercial processes for generating hydrogen, SMR is currently the most economically attractive due to its high efficiency and low feedstock, production, and operational costs. However, SMR is a significant source of greenhouse gas (GHG) emissions, emitting about 9 kg of CO_2 per kg H_2 produced. The GHG emissions are primarily a result of the use of natural gas combustion to supply the heat needed in SMR furnaces to drive the chemistry to high conversion. Therefore, in order for the hydrogen generation sector to significantly reduce its share of GHG emissions, the rapid development and deployment of sustainable low-carbon energy solutions are warranted.

This study examines a potential decarbonization solution: electrification of SMR. We use a bench-scale electrified reformer that uses resistive heating as the heat source by applying direct electric current through a coiled high-resistance FeCrAl metal wire. The metal coils (0.5-mm o.d.; 40 cm in length; 3.13 ohm resistance) used were uncoated or coated with a Ni/zirconia catalyst. A feed stream comprised of steam and methane (4:1 molar ratio) and nitrogen at 190°C and 1 atm flows into a stainless-steel tube (45 mm i.d.) that houses the wire. The feed perpendicularly encounters the electrically-heated wire 100 mm downstream. Experiments are conducted at power settings in the range 50-90 W, corresponding to an estimated wire temperature range of 650-850°C. After condensing unreacted steam from the effluent exiting the reformer, FTIR analysis is used to detect and quantify for CO, CO_2, and CH_4.

Results for the coated coil showed that power settings below 66 W yield low conversions (< 3.0%). However, for power greater than 66 W, an unusual step change in conversion occurred – 84% at 70 W increasing to 89% at 89 W. The ratio of CO to CO_2 increased from 0.77 at 58 W to 1.7 at 89 W. Experiments with the uncoated coil exhibited insignificant conversions (< 0.5%) between 50-90 W. This clearly demonstrated that the high conversions obtained were due to the catalytic activity of the coating. Though H_2 was not measured directly, we computed H_2 generation from the conversion, CO/CO_2 composition, and SMR chemistry – at the highest conversion, we obtained a value of 0.16 mol of H_2 generated per mol of feed mixture. This corresponds to USD 93 per kg H_2.

Our results show that under the experimental conditions examined in our study, electrification of SMR, albeit expensive in terms of H_2 production cost, is possible with high conversions. This will have significant relevance to the hydrogen generation sector as it makes attempts to reduce its carbon footprint. Further studies are ongoing to investigate ways to reduce the H_2 production cost and to gain an understanding of the step change behavior in conversion.

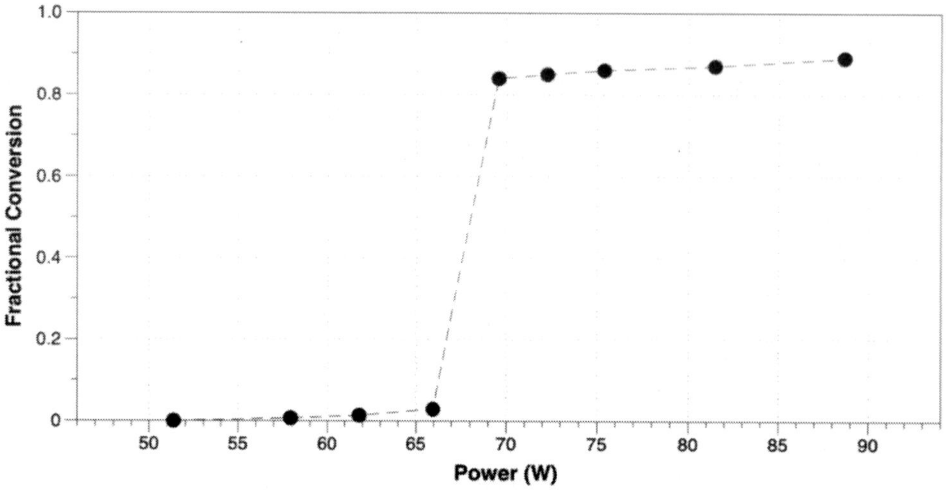

Figure 1: Fractional conversion of methane as a function of applied power.

222h A Novel Electric Reactor Concept for Methane to Acetylene Conversion

Shauvik De, Alexander P. van Bavel

Abstract: With renewable electricity becoming a highly widespread and accessible form of energy, electrification of chemical processes presents one of the most promising transition paths to low carbon footprint manufacturing of chemicals. A large extent of chemical processes in industry are endothermic in nature and the necessary heat is currently provided by burning fossil fuels. The conversion of methane into acetylene has gained more attention in recent years and is one of the promising pathways to create acetylene. This conversion is favorable at high temperature (>1900 ^{0}C) at very low residence times (<1 millisec), which poses some challenges. The basic technology using an electric arc reactor, however, has been around for many decades [1]. Due to heterogeneity of the temperature and residence times, significant amounts of carbon is formed, which decreases the selectivity towards acetylene in conventional reactor concepts. The current work shows a novel reactor concept where necessary heat is provided directly via electricity (joule heating) for a high temperature chemical conversion of methane to acetylene [2]. The concept involves flow of methane through structured packing or obstacles of arranged mesh or wire structures directly heated by green electricity [fig.1]. Such concepts are also very recently published in literature for catalytic conversions [3-5]. However, a detailed computational fluid dynamics (CFD) based numerical simulations coupled with developed reaction kinetics shows optimal conversion and selectivity of methane to acetylene at the desired temperature and residence times, if these structures are arranged in a specific configuration. The concept also shows added advantage with respect to scale up, modularity, residence time distribution and lower pressure drop than conventional reactor concepts.

1. High-Temperature Pyrolysis of Hydrocarbons. 1. Methane to Acetylene Industrial & Engineering Chemistry Process Design and Development, A. Holmen, O. A. Rokstad, and A. Solbakken, 1976 15 (3), 439-444
2. A method of producing C2H2 –De Shauvik, P. van Bavel Alexander, European Patent; EP 3 845 513, 2020
3. Enhancing CO_2methanation over a metal foam structured catalyst by electric internal heating - Dou L, Yan C, Zhong L, Zhang D, Zhang J, Li X, Xiao L - Chem Commun, 2019
4. A compact catalytic foam reactor for decomposition of ammonia by Joule heating mechanism- A Badakhsh, Y Kwak, YJ Lee, H Jeong, Y Kim, H Sohn - Chemical Engineering Journal, 2021
5. Electrified methane reforming: A compact approach to greener industrial hydrogen production - Wismann ST, Engbæk JS, Vendelbo SB, Bendixen FB, Eriksen WL, Aasberg-Petersen K, Frandsen C, Chorkendorff I, Mortensen PM. Science, 2019.

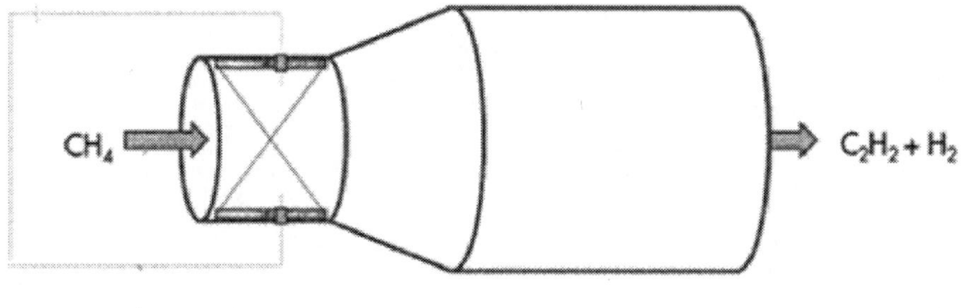

391a Electrification of Distillation for Decarbonization

Chengtian Cui, Yufei Zhao, Cornelius Masuku

Distillation is one of the largest energy consumers in the chemical industry. Currently, distillation columns are largely driven by utility steam generated by the combustion of fossil fuels. To decarbonize distillation columns, heat pumps have become an effective way. Heat pump is usually electrically-driven so it can be integrated with renewable power. In this work, we have designed several distillation configurations to separate chemical components with either azeotropic or zeotropic properties. The implementation of electrical equipment also brings additional economic costs. Therefore, the evaluation of additional capital cost versus additional saving in operating cost should be considered. In addition, the use of heat pumps can generate additional process control issues, thus we compared the dynamics and control performances of both thermal- and electrical-driven processes. For specific systems, the electrical distillation processes have significant economic advantages over the conventional thermal-driven ones. Also, the complicated electrified processes can be well controlled when facing large throughput or composition disturbance.

391b Microwave-Driven Distillation Separation Technology - Full of Breakthrough Opportunities in Electrified Chemical Processes

Kai Liu, Zhenyu Zhao, Hong Li, Xin Gao

Microwave irradiation with selective heating characteristics has been confirmed to affect the relative volatility of binary mixture, which should push the development of novel separation technology based on the difference of dielectric property of the composition in chemical engineering. The application of newly electrified microwave heating to traditional processes of vapor-liquid mass transfer has given rise to many emerging process intensification technologies, such as microwave-induced evaporation separation and microwave-induced reactive (flash) distillation. In this presentation, we summarize the role of microwaves in the above processes from the driving power and microwave special effects; and briefly outline the bottlenecks in the current development to point out the research direction for subsequent fundamental investigations and industrial applications.

For modeling and design the separation process of microwave induced evaporation. A novel dimensionless number Z_{MW} was derived in the present study based on an assumption involving molecular radiators to explore the effect of dielectric and thermodynamic properties of the materials as well as the MW field intensity on the microwave-induced relative volatility change (MIRVC). Moreover, a quantitative correlation was established between MIRVC and Z_{MW}, whose model parameters were determined by fitting experimental data obtained from mixtures under MW irradiation. The correlation was also employed to predict MIRVC in data obtained from previous studies. The error range between the predicted and experimental values was within ±6%, indicating the validity of the proposed quantitative correlation.

Furthermore, a novel MW-VLE model was proposed to predict the VLE data of binary systems in microwave fields using system thermodynamic properties, dielectric properties, and electric field strength. For modeling, first, the influence of thermodynamic and dielectric properties on microwave-induced relative volatility change (MIRVC) was investigated experimentally in detail. It indicated that increasing the system dielectric loss difference, boiling temperature difference, and decreasing the thermal conductivity can enhance the MIRVC. Then, model characteristics parameters were obtained by fitting the VLE curves with/without microwave radiation. Afterward, the empirical correlation equations of MW-VLE were derived by correlating model characteristics parameters with mentioned above factors and electric field strength.

Finally, an innovative microwave-induced spray evaporator was developed to overcome the current defects. Furthermore, the experiments demonstrated an excellent separation efficiency of alcohol-water azeotropic systems are achieved via microwave-induced spray evaporator. Moreover, the experiments of several mixtures with different permittivity and dielectric loss were performed to elucidate the mechanism of microwave-induced relative volatility change (MIRVC). The experimental results revealed that the increase of high dielectric loss components in the vapor phase under microwave heating in comparative experiments verifies the inference that dielectric loss determines the direction of MIRVC.

391c Microwave Regeneration of Ionic Liquid Sorbents for Direct Air Capture: Finite Element Modelling and Experimental Demonstration

Aidan Klemm, Eda Cagli, Muhammad Zeeshan, Burcu Gurkan

As the world strives toward carbon neutrality to stave off an impending global climate disaster, decreasing reliance on fossil fuels is an essential step to reaching emission mitigation targets. Significant research efforts are being made toward electrifying industrial processes, reducing dependence on combustion for heating and energy supply. Direct air capture (DAC) of CO_2 is no exception. It is a key area in which electrification will drastically improve carbon capture efficiency of existing technologies. Reactive absorption of CO_2 is essential to reach the required selectivities for effective DAC, which has a feed of only 0.04% CO_2. This reaction is exergonic, and thus energy is required to reverse it and regenerate the sorbent for cycling. Supplying this energy via microwaves is a promising method for regeneration as many DAC materials react with CO_2 to form charged functionalities, which respond more intensely to microwave irradiation. In this study, we present the successful microwave-assisted regeneration of a CO_2-reactive ionic liquid (IL), 1-ethyl-3-methylimidazolium 2-cyanopyrrolide, which has been shown to have excellent CO_2 capacities (~1 mol CO_2/kg sorbent) at DAC relevant conditions. We further incorporated this IL into a metal organic framework (MOF), specifically zinc imidazolate framework 8 (ZIF-8), to form a powdery composite sorbent, which is assessed via breakthrough analysis for both absorption and desorption and under humid conditions. To model the microwave regeneration of these sorbents and assess energy utilization, the Maxwell equations were coupled with the Navier-Stokes transport equations and solved via finite element modeling. We find that the distribution of the electromagnetic energy is highly dependent on sample geometry and the dielectric properties of the sample. These results provide key insight into the relevant parameters for improving energy efficiency in microwave-based separations, and will inform the geometries and measurement techniques for future studies.

391d Energy Efficient Approach for Direct Air Capture

Rohan Sartape, Rohit Chauhan, Meenesh R Singh

Accomplishing the Department of Energy's goal of capturing CO_2 from the air for less than \$100/metric ton of CO_2 equivalents warrants an urgent development of energy-efficient capture technologies. Out of the technologies available for CO_2 capture, water-driven CO_2 capture is attractive because they incur a low energy penalty. The Migration Assisted Moisture Gradient (MAMG) process captures and regenerates CO_2 based on the bicarbonate-carbonate equilibria. The CO_2 capture occurs in an organic phase and captured CO_2 migrates to the aqueous phase separated by the membrane under an external bias. Ionic Liquids (ILs) can improve the energy efficiency of the process due to their low vapor pressure and high conductivities achieved by hydrogen bonding. Herein we present the results of improving the energy efficiency of the MAMG process by studying the effects of various ILs. Moreover, we also demonstrate the results from the large-scale operation of the MAMG process in a Bipolar Membrane Electro Dialysis unit (BPMED) unit for CO_2 capture from air. The BPMED consists of alternating layers of BPM and AEM separating the aqueous and the organic medium. The water-splitting reaction at the interface of the BPM additionally maintains electroneutrality throughout the organic and aqueous phases. Parameters such as the surface area of the membrane, effects of IL concentration, varying composition of CO_2, impurities such as SO_x, and relative humidity of the air on the rate of CO_2 capture will also be presented.

391e Electrochemical Direct Air Capture Using Neutral Red As an Oxygen Insensitive Redox-Active Molecule: System Design and Performance

Hyowon Seo, Michael Massen-Hane, Timothy Nguyen, T Alan Hatton

Carbon capture and storage is a collection of techniques that capture CO_2 from large emission sources or from the atmosphere and safely stores it underground or in products. Carbon capture and storage is a versatile technology that enables both emission mitigation from industry as well as CO_2 removal via direct air capture.

The electrochemical direct air capture using organic dye compound, neutral red/leuco-neutral red (NR/NRH2), as an oxygen insensitive redox-active molecule uses electrochemical potentials to modulate pH of water to achieve CO_2 separation from ambient air. A preliminary system was assembled to demonstrate separation using 15% CO_2 and ambient air in a batch and continuous flow system, designed to evaluate carbon capture efficiency and system integrity.

To obtain better energy analysis, a cyclic flow system was built to study the performance at different current densities, liquid flow rates, electrochemical cell design, and electrode materials, using 15% CO_2. The current and voltage responses of the electrochemical process in a cyclic flow setup will be discussed under various electrochemical conditions. The integrated electrochemical system under the optimized condition was tested for direct air capture in a cyclic flow system.

The demonstrated electrochemical direct air capture system using NR/NRH2 at high electron utilization and moderate electrical energy consumption will encourage future research into optimizing the electrochemical separation unit for long-term and stable operations in large-scale applications.

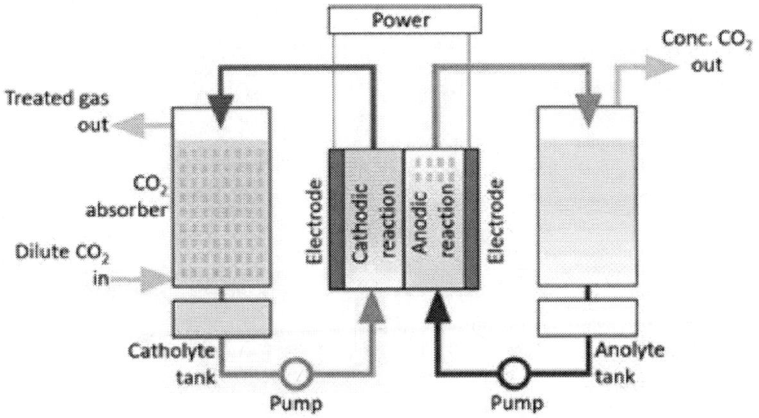

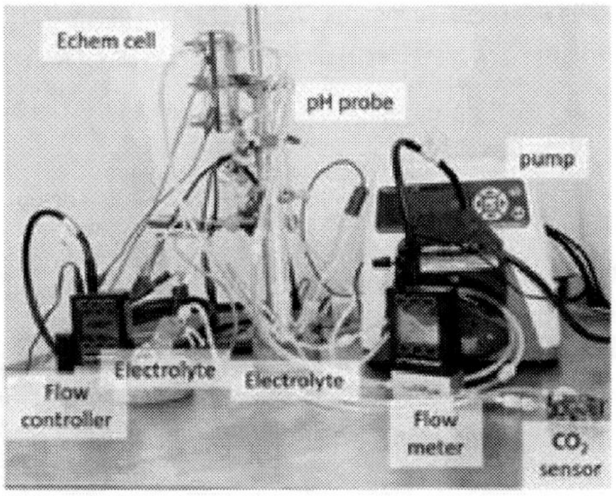

391f Selective Electrochemical Separation of Cations: Rare Earth Element Recovery and Water Softening

Haley Vapnik, Johannes Elbert, Shao-Wei Tsai, Xiao Su

Rare earth elements (REEs) have become critical resources in various electronic devices and renewable energy technology; due to their unique magnetic, phosphorescent, and catalytic properties. There are currently several ways to separate REEs, including chemical precipitation, solvent extraction, ion exchange, and adsorption; however, these methods are highly inefficient, produce significant waste, and produce low-purity REEs. Ion removal using electrochemical separation methods is highly selective and green. However, these methods are typically used for anion adsorption.

Electrochemical techniques can be used to capture and release REEs reversibly using redox-copolymers. Combining the functionalities of ion-exchange groups, such as carboxylic acids, for REE adsorption, and redox groups, such as ferrocene moieties that can be controlled electrochemically for regeneration/desorption results in a redox switchable copolymer. We observed an yttrium (Y) adsorption capacity of 69.4 mg/g polymer, and electrochemical regeneration of these electrodes yielded a regeneration efficiency close to 100%, as well as similar uptake and recovery for other rare earth elements, including cerium (Ce), neodymium (Nd), europium (Eu), gadolinium (Gd), and dysprosium (Dy), establishing the concept of electrochemically regenerable ion-exchange copolymers for REE recovery. Thus, electrochemical controls can be used in REE recovery systems, providing highly effective methods that provide efficient recovery of REEs without the use of chemical reagents while also offering high modularity and sustainability.

Finally, redox-copolymers have also been utilized for electroseparating other cations, not just REEs, such as alkaline-earth cations for water softening. Various technologies can be used to remove hard ions from water, including ion exchange, chelation, distillation, and reverse osmosis. We demonstrate 100% selectivity towards binding divalent hard ions over monovalent competing ions by the redox-copolymer. By electrochemically mediated regeneration, hard ions can be removed from sorbent materials and regenerated without any regenerative chemicals needed for multiple cycles.

391g A Novel Approach for Direct and Continuous Electrochemical Manufacturing of Lithium Hydroxide from Geothermal Brines

Yuge Feng, Sibani Biswal

The increasing demand for lithium in lithium-ion batteries (LIBs) over recent decades has necessitated the development of efficient and sustainable processing of lithium resources. Geothermal brines contain large amounts and high concentration (up to 3800 mg/L[9]) of lithium, making them an attractive and feasible source of lithium. However, high-purity lithium extraction from brines remains a significant challenge due to the presence of a large number of unfavorable ions, such as sodium ions (Na^+).

Herein, we present an effective and sustainable strategy for electrochemical lithium recovery from geothermal brines. Our approach utilizes a three-chamber solid electrolyte (SE) electrochemical reactor with a lithium conductive ceramic selective membrane (LISICON). As depicted in Fig 1, the simulated brine containing LiCl and NaCl flows into the middle layer filled with SE, while water flows in both the anode and cathode. The functionalized SE with sulfonate groups guarantees efficient Li^+ and Na^+ conduction, while only Li^+ could transport through the lithium-ion conductive ceramic membrane, and combine with the OH^- generated from HER. Starting from the simulated brine with initial Li: Na ratio of 1:100, we achieved a high Faradaic Efficiency (FE) of 97.51 ± 3.2%, based on Li^+ concentration, and a Li^+/Na^+ selectivity ratio of over 2×10^3. Our device continuously recovers LiOH from brine with less contamination of Cl_2 and HClO, and the average cell voltage corresponds to energy consumption of 9.87 kWh/kg_{Li}, which represents a carbon footprint of 8.39 $kgCO_2e/kg_{LiOH}$. The presence of Na^+ would lead to an increase of cell voltage and partially unreversible membrane fouling, but the ceramic membranes can be regenerated partially by operating with LiCl (1 M) solution in the middle layer. With ion chromatography (IC) analysis of the catholyte and Time-of-Flight Secondary Ion Mass Spectrometry (TOF-SIMS) analysis on the ceramic membrane, we confirmed that lithium is successfully recovered only from the simulated brine.

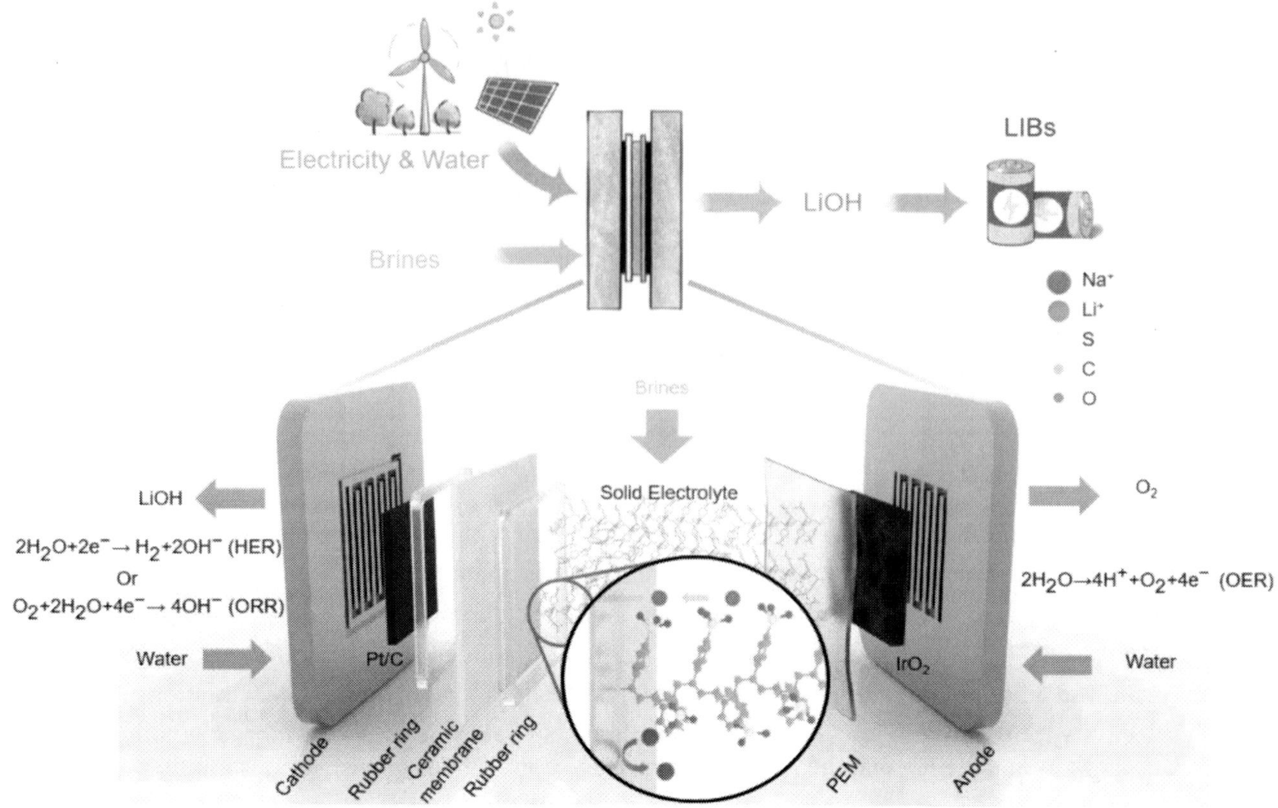

Electricity & Water
Brines
LiOH
LIBs
Na⁺
Li⁺
S
C
O
Brines
Solid Electrolyte
LiOH
2H2O+2e⁻→H2+2OH⁻ (HER)
Or
O2+2H2O+4e⁻→4OH⁻ (ORR)
Water
Pt/C
O2
2H2O→4H⁺+O2+4e⁻ (OER)
IrO2
Water
Cathode
Rubber ring
Ceramic membrane
Rubber ring
PEM
Anode

391h Heat Pumps for Electrification of Multicomponent Distillation Trains: An MINLP Formulation

Akash Nogaja, Mohit Tawarmalani, Rakesh Agrawal

Thermal separations are crucial to chemical and petrochemical sectors, but they significantly contribute to fossil fuel consumption and greenhouse gas emissions. Distillation is the primary separation technology, accounting for 90-95% of all liquid phase separations[1] and around 2.5% of total energy consumption in the USA[2]. To address this sustainability challenge, it is critical to enhance the energy efficiency of distillation.

One potential solution is to electrify the process using heat pumps. This will reduce greenhouse gas emissions by utilizing energy from renewable sources while increasing the exergy efficiency of distillation, ultimately reducing the energy required for the given separation[3]. However, feasibility and efficiency of heat pumping for multi-component separation depends on the operating conditions and sequencing of splits. A sequential optimization approach that involves selecting configuration before integrating heat may produce sub-optimal designs and leave valuable opportunities untapped. Therefore, an approach that models and globally optimizes distillation train sequencing is needed.

In this study, we present a new shortcut model, that relies on a simple algebraic equation which accurately predicts the work requirement of a mechanical vapor recompression type heat pump for any number of components without using cumbersome thermodynamics. We integrate this equation into a superstructure based Mixed Integer Non-Linear Programming (MINLP) formulation which embeds the complete search space of configurations, thereby creating the first comprehensive approach to finding globally optimum Heat Pump Assisted Distillation (HPAD) designs for general `n' component zeotropic separations. Using a few case studies, we demonstrate the potential savings and greenhouse gas reductions that can be achieved by exploring the use of heat pumps.

In summary, this study highlights the significant role of heat pumps in improving energy efficiency of distillation while reducing greenhouse gas emissions. The proposed approach offers a promising pathway towards a more sustainable future for chemical and petrochemical industry.

(1) Humphrey, J. L.; Keller II, G. E. Separation Process Technology: Performance, Selection, Scaleup. *McGraw Hill* **1997**.

(2) Chapas, R. B.; Colwell, J. A. Industrial Technologies Program Research Plan for Energy-Intensive Process Industries. **2007**. https://doi.org/10.2172/1218715.

(3) Tumbalam Gooty, R.; Chavez Velasco, J. A.; Agrawal, R. Methods to Assess Numerous Distillation Schemes for Binary Mixtures. *Chemical Engineering Research and Design* **2021**, *172*, 1–20. https://doi.org/10.1016/J.CHERD.2021.05.022.

432a Systematic Assessment of Alternative Low-Carbon Ethylene Production Pathways

Guiyan Zang, Dharik Mallapragada

Widely used as an essential building block for making resins and plastics, ethylene is a key platform chemical with a relatively large carbon footprint and production concentrated in the United States, Middle East, and China. Today, ethylene is manufactured through steam-cracking of ethane and naphtha, which has CO_2 emissions of 1~1.6 tonne (t) per t of ethylene. As material demand for ethylene is anticipated to increase with economic growth in various regions, there is an urgency to develop and deploy sustainable production routes and avoid "emissions lock-in" from the deployment of existing production routes. Possible options under consideration include: a) direct electrification to substitute fossil fuel for heating during steam cracking, b) use of low-carbon hydrogen to provide heat for the cracking process, where the hydrogen is produced via alternative approaches including byproduct methane reforming with carbon capture and storage (CCS), c) dehydration of ethanol produced from sustainably available biomass resources as well as d) less mature, but potentially transformative approaches based on electrochemical conversion of ethane or methane. These technological options span alternative feedstocks, novel interactions with the other energy infrastructure, like the electricity grid and hydrogen infrastructure, and have different scaling properties, which must be considered when performing a comparative analysis.

The objective of this work is to perform a systematic techno-economic and environmental analysis of alternative low-carbon ethylene production routes that consider the above-mentioned factors. Our approach is based on: a) developing detailed process simulations for each process to allow for characterizing mass and energy balances and unit sizing, b) techno-economic analysis (TEA) to compare the cost and economic risks of each process according to various types of energy system interactions and c) life cycle analysis (LCA) to quantify the greenhouse gas (GHG) emissions impact spanning cradle-to-gate system boundary, that considers co-product disposition and energy infrastructure interactions. The results will provide insights for research and development teams and industry stakeholders by identifying key process drivers that will help reduce costs, emissions, and investment risks for each process.

432c Optimal Flowsheets for Electrification of Oil Refineries through Superstructure Optimization

Sampriti Chattopadhyay, Ana I. Torres, Ignacio Grossmann

The chemical industry is investing in renewable resources and exploring various initiatives to achieve energy transition and decarbonization goals, such as using green hydrogen, carbon capture and storage technologies, electrification of heat, and electrochemical conversion. However, uncertainties around new technologies, electricity and carbon prices, and capital investment make navigating the transition complex. The industry must change to meet decarbonization standards, but valuable infrastructure should not be prematurely discarded for emerging technologies. There is a dilemma of when to transition and which initiative to adopt, as switching too early leads to inefficiencies and switching too late increases the carbon footprint.

In this work we address the problem of determining the optimal retrofit electrification of an existing oil refinery. We propose a generic process refinery flowsheet that represents the current operations of an oil refinery. We then consider several electrification alternatives that have been identified from the literature. These electrification alternatives together with the refinery flowsheet are included in the superstructure that is to be optimized. Examples of the alternative electrification technologies include e-boilers, alternative pathways for hydrogen generation, connection to decarbonized electricity grids and battery storage. The different electrification alternatives are combined with the initial flowsheet to create a network superstructure of alternative flowsheets in which existing units of the refinery can be removed, replaced by new units, or reassigned to perform a different task, which in turn may require new interconnections of. the units. Data on the technological readiness level, capital and operating costs, and uncertainties associated with each alternative have been obtained from various sources.

The optimization of the superstructure to find the minimum cost electrified refinery is formulated as a Mixed-integer Linear programming model that includes simplified process equations, as well as logical and feasibility constraints for removal and addition of units and interconnections. The proposed objective function includes capital and operating expenses of the potential technologies to be selected, as well as taxes for CO_2 emissions to promote sustainability goals. This MILP model is also extended to a multi-period optimization problem in order to consider long-term planning horizons, e.g. a ten-year long plan for a given forecast of product demands, and that can determine the optimal timeline for switching to the decarbonized options. Numerical results of the proposed MILP models are presented for nominal values of the parameters, as well as for several scenarios that include changes in the costs and performance of electrification technologies, the price of crude oil, and the carbon tax. The results show the potential economic advantages associated the optimal electrification for the retrofit of oil refineries. Finally, we briefly discuss how the proposed MILP models could be extended with stochastic programming techniques to explicitly account for uncertainties in the parameters of the MILP models.

432d Critical Assessment on Indirect Ocean Capture of CO_2 for Desalination Brine Valorization

Yian Chen, Xiaowen Chen, Eric Tan, Jacob S. Kruger, Robert Nelson, Jeffrey Linger, Violeta Sànchez i Nogué, Ling Tao, Yoram Cohen, Gregg Beckham, Joshua A. Schaidle

Given the concerns of global climate change, the concept of CO_2 removal from the atmosphere has attracted substantial interest. Among the various CO_2 removal technologies, indirect ocean capture (IOC) is an emerging approach for direct removal of atmospheric CO_2 via extraction of dissolved inorganic carbon (DIC) from seawater, followed by air-seawater re-equilibration. Unlike CO_2 capture from large stationary emission sources, IOC has the potential to manage both the existing excess CO_2 in the atmosphere, as well as CO_2 from mobile or smaller point sources. Moreover, given that the CO_2 concentration (mg/L) in seawater is approximately 140 times greater than that in air, IOC can avoid processing the large volumes of air (e.g., via pumping and compressing) that is required for Direct Air Capture (DAC) techniques. Motivated by the growing body of literature on IOC in the last 5 years, in this study, we present a critical review and provide a quantitative perspective on IOC technologies which then provides support for the proposed process scheme for integration of IOC with a seawater reverse osmosis (SWRO) desalination plant to extract CO_2 from RO brine. Techno-economic analysis (TEA) and life-cycle assessment (LCA) was then carried out to evaluate the mutual benefits to both the IOC facility and the desalination plant via.

Here we note that control of solution pH and CO_2 partial pressure has been employed in various proposed IOC technologies, including for example, membrane separation, ion-exchange resin, and electrochemical techniques. However, most of the studies are still at the bench-scale proof-of-concept stages and face various challenges that include: (i) insoluble scales form and deposit at the cathode and membrane surfaces, leading to increased cell resistance and compromised electrolytic efficiency, (ii) slow CO_2 extraction rate and thus low extraction efficiency, (iii) high electrochemical energy consumption, and (iv) high capital and operating costs due to membrane contactor vacuum stripping and seawater pretreatment. At the current state-of-the-art, bipolar membrane electrodialysis (BPCEM) has been touted as one of the most practical IOC methods (with up to 98% lower electrochemical energy consumption). The above technology was thus considered in analyzing the potential adoption of IOC via TEA and LCA.

TEA analysis suggests that a stand-alone IOC facility is not yet feasible. However, the overall IOC cost via co-location with other ocean water-pumping technologies, especially seawater desalination plants, can be reduced up to 59% compared to a stand-alone facility. The IOC cost can be further reduced by utilizing SWRO desalination brine, which will double the CO_2 concentration in the seawater; hence, a reduction in the volume of seawater that requires processing via IOC. More importantly, as CO_2 solubility decreases with increasing solution salinity, a much higher CO_2 extraction rate can be achieved by spontaneous degassing of the acidified RO desalination brine (73%) relative to acidified seawater (40%). Consequently, the proposed IOC process, co-located with a seawater desalination plant using RO brine for CO_2 extraction, can reduce the cost of IOC to \$138/tCO_2, which is ca. 96% lower than a stand-alone IOC system. It is also stressed that IOC co-location is beneficial to the seawater desalination plant since it will lead to (i) generation of extra revenue through RO brine valorization, and (ii) lowering the total volume and salinity of the brine that needs to be discharged for disposal, ultimately mitigating the negative environmental impact of the desalination process. In addition, the carbon credit earned through CO_2 capture can provide extra revenue for the desalination industry, leading to a reduction of freshwater production cost by ~6% (using a projected carbon price of \$150/tCO_2 in 2050). Finally, integrating IOC can help the desalination industry achieve negative carbon emissions. In conclusion, the IOC co-location with seawater desalination plants will have the benefits of economic CO_2 removal, combating global climate change together with other CO_2 removal technologies, and improving the sustainability and cost efficiency of seawater desalination. Implementing the above proposed approach will require research efforts that focus on the development and optimization of the IOC technologies.

432e Modeling and Simulation of Electrodialysis Processes for Nutrient Recovery from Wastewater

Dylan Weber, Joseph K. Scott

Electrification of the fertilizer production industry can enable serious decarbonization efforts globally, especially those based on renewable electricity. Currently, ammonia is manufactured based on the heavily centralized Haber-Bosch process, which accounts for 2% of global energy consumption and 1.2% of carbon emissions. Moreover, phosphorous mines create heavy metal hazardous waste streams and are ultimately non-renewable. As an alternative, recovery of fertilizer nutrients from wastewater streams, either in wastewater treatment plants (WWTPs) or concentrated animal feeding operations (CAFOs), has been proposed to achieve more decentralized and sustainable fertilizer production. These wastewater streams contain large amounts of ammonia (N) and phosphorus (P) that can be recycled for direct use on farms. Currently, these components are not recovered in the vast majority of WWTPs and CAFOs. Usually the N/P components are either neutralized or directly released into the environment. The challenges with recovering these products is that the starting streams are very dilute, highly variable, and geographically distributed. Therefore, a viable recovery process must be flexible, modular, and able to exploit distributed renewable energy sources.

Currently, there are many methods for nutrient recovery including crystallization, air stripping, and biological treatment. However, these methods require additional chemicals, operate at high pH, require careful biological controls, and are not well-suited to modular implementations using renewable electricity. In contrast, electrochemical separations may provide a very promising route for distributed nutrient recovery . Over the last few decades, electrochemical separations have been used extensively for desalination purposes to meet drinking water needs where resources are constrained. Specifically, electrodialytic (ED) ion-exchange membrane-based processes allow for lower energy consumption and direct energy usage in the form of renewable electricity versus thermal or pressure driven systems. Electrodialysis systems also have fast dynamics which can easily deal with fluctuations in power caused by using renewable energy intermittency. Given the vast number of commercial installations for desalination, there is scant literature published on electrochemical separations for nutrient recovery of fertilizer from WWTP or CAFO streams.

This talk will present a novel modeling framework and process simulation tool for ED-based nutrient recovery processes. The tool can predict pilot and commercial scale performance to quickly iterate system designs based on number of stages required, membrane area, as well as energy requirements. This simulation tool also considers the key dynamics caused by renewable energy and feedstock variability, and is therefore well suited for studying process control strategies. This is a significant contribution because, at present, commercial chemical process simulators do not support electrochemical unit operations such as ED, and there is no easily accessible or standardized simulation code for such processes in the literature. Using this tool, we can pave a way for future process control strategies and design modular processes that can be dispatched where needed.

432f Optimizing the Transition to Sustainable Ammonia Fertilizer Supply Chains

Matthew Palys, Prodromos Daoutidis

Renewable ammonia production using hydrogen sourced from electrolysis is an emerging approach to alleviating the 2% of global emissions attributed to synthetic nitrogen fertilizer production. A distributed implementation of renewable production facilities close to the agricultural end user has additional potential to mitigate the costs and emissions associated with interstate or even international transport of ammonia currently produced at large-scale centralized facilities. Local ownership of new renewable ammonia production could insulate the end user (e.g., the farmers) against fluctuations in fertilizer prices that have historically occurred on the global ammonia market. In the United States, recent federal policy such as investment and/or production tax credits for power generation and hydrogen production with low associated carbon intensity [1] has improved the economic viability of this approach to decarbonizing agriculture.

Despite the potential benefits, replacing ammonia fertilizer produced using fossil fuels and transported over a national or even global scale with such locally produced renewable ammonia will not be instantaneous. To this end, we develop a multi-scale optimal supply chain transition model which maximizes the economic benefit of developing a local renewable ammonia supply chain. We first optimize renewable ammonia production facility to minimize the levelized cost of ammonia over a range of (i) installation years from present-day to 2030 which affects cost curves of constituent technologies such as electrolysis, (ii) production scales, and locations with differing wind and solar renewable generation potential. Renewable variability is accounted for through a combination of intermediate storage (e.g., batteries, hydrogen buffer) and oversized chemical production pursuing variable production rates. These design features are optimized through simultaneous technology selection, sizing, and hourly scheduling with hourly resolution representative years for wind and PV capacity factors in each location used as input data, as is the prevailing approach in the literature [2-4]. The results of this production facility optimization are subsequently used as input data to a supply chain transition optimization model. This model has annual resolution and determines the installation year, production scale, and location of new renewable ammonia production facilities as well as how this new means of ammonia production will augment and eventually replace importing ammonia from the existing producers in meeting spatially resolved (e.g., county-by-county) demands [5] in a manner which minimizes the 30 year net present cost of the supply chain.

We present a case study for Minnesota's ammonia fertilizer supply chain transition demonstrating the economic viability of such a transition by 2032 over a range of historically observed market ammonia prices. This analysis includes constraints on the scale of renewable generation facilities informed by existing installations in Minnesota as well as electrolysis availability informed by global projections of annual manufacturing. The analysis also incorporates the possibility of claiming investment or production tax credits for renewable power generation and clean hydrogen production to understand how such policy incentivizes the adoption of decarbonized fertilizer production and to quantify the economic opportunity created by such policy from the perspective of the state of Minnesota. Ultimately, we hypothesis that such an economically optimal renewable ammonia supply chain transition will enable significant reduction in the carbon intensity of fertilizer supply along with considerable opportunities for local economic development.

References

[1] Inflation Reduction Act of 2022, H.R.5376 (2022). https://www.congress.gov/bill/117th-congress/house-bill/5376/text

[2] Fasihi et al. (2021). Global potential of green ammonia based on hybrid PV-wind power plants. *Applied Energy*, *294*, 116170.

[3] Bose et al. (2022). Spatial variation in cost of electricity-driven continuous ammonia production in the United States. *ACS Sustainable Chemistry & Engineering*, *10*(24), 7862-7872.

[4] Palys & Daoutidis (2022). Power-to-X: A review and perspective. *Computers & Chemical Engineering*, 107948.

[5] Palys et al. (2018). Exploring the benefits of modular renewable-powered ammonia production: A supply chain optimization study. *Industrial & Engineering Chemistry Research*, *58*(15), 5898-5908.

432g Graph-Based Optimization for Technology Pathway Analysis: A Case Study in the Decarbonization of University Campuses

Blake Lopez, Jiaze Ma, Victor Zavala

The global mean surface temperature reached 1°C above pre-industrial times in 2017 and is projected to reach 1.5°C as early as 2030 [1]. Significant efforts need to be made to decrease greenhouse emissions to limit the impacts such as biodiversity loss and increased severity/frequency of severe weather events [2,3]. Large industrial sectors such as urban centers, chemical companies, manufacturing facilities, and microgrid systems are actively looking for technology pathways that can help minimize their carbon footprint. An important and representative set of industrial systems that is actively seeking to decarbonize operations are university campuses. Campuses involve large collections of buildings that consume heating, cooling, power, and transportation services. To give some perspective into the magnitude of these systems, we note that over 16.5 million students in the US reside on university campuses [4]. It is estimated US campuses emit 7.7 MTCO2e per student and are responsible for 2% of the 5,222 million MTCO2e of the US greenhouse gas emissions [5,6].

The number of new technologies that can be used to decrease greenhouse gas emissions is steadily increasing and complicates decision-making processes. In the context of university campuses, for instance, detailed mixed-integer optimization models have been developed to determine suitable system configurations [7-10]. Existing optimization models are typically intended to provide long-term investment plans to achieve decarbonization and are often hindered by the fact that there might not be sufficient data to evaluate emerging technologies.

In this work, we present a graph-based optimization framework for analyzing technology pathways that can provide desired product demand targets while minimizing supply and technology costs by seeking to maximize the total surplus. Our modeling approach aims to provide a unifying framework that captures diverse technologies and their interconnectivity via products. Moreover, the framework aims to use a minimum amount of data for technologies (e.g., efficiencies, capacities, operating costs, investment costs), thus providing a "high-level" picture on the technology landscape and on potential factors that influence the economic viability of diverse technologies. Moreover, this approach aims to enable fast screening of pathways and to conduct parametric studies under diverse market and policy scenarios.

We show that the proposed graph-based model can be interpreted as a value chain in which there is no transportation and spatial context; specifically, the model is a simplification of a multi-product supply chain model [11,12]. This observation is important, as the model proposed can provide valuable preliminary insights that can inform the development of more sophisticated infrastructure models (e.g., that capture transportation and spatial effects). Duality analysis of the proposed model reveals that this has a natural market (value chain) interpretation that can help identify technology pathways that maximize profit for stakeholders and that helps reveal the inherent value (prices) of intermediate products. This price information is key in market studies that involve sector coupling (e.g., hydrogen, ammonia, and power). Moreover, the graph-based abstraction reveals how externalities (e.g., carbon taxes or product demands) propagate through the value chain and affect prices of diverse products.

Our developments are illustrated via a case study involving a prototypical university campus that seeks to identify pathways to reduce its carbon footprint (e.g., via electrification and deployment of hydrogen technologies). Analysis conducted with our framework reveals carbon tax values, technology targets, and investment budgets that can help achieve different levels of decarbonization.

[1] V. Masson-Delmotte *et al.*, "An IPCC Special Report on the impacts of global warming of 1.5°C above pre-industrial levels and related global greenhouse gas emission pathways, in the context of strengthening the global response to the threat of climate change, sustainable development, and efforts to eradicate poverty," p. 630.

[2] O. Hoegh-Guldberg *et al.*, "The human imperative of stabilizing global climate change at 1.5°C," *Science*, Sep. 2019, doi: 10.1126/science.aaw6974.

[3] D. A. Smale *et al.*, "Marine heatwaves threaten global biodiversity and the provision of ecosystem services," *Nat. Clim. Chang.*, vol. 9, no. 4, Art. no. 4, Apr. 2019, doi: 10.1038/s41558-019-0412-1.

[4] "NCES Digest of Education Statistics. URL: nces.ed.gov." https://nces.ed.gov/programs/digest/d16/tables/dt16_307.10.asp (accessed Apr. 19, 2022).

[5] P. Sinha, W. A. Schew, A. Sawant, K. J. Kolwaite, and S. A. Strode, "Greenhouse Gas Emissions from U.S. Institutions of Higher Education," *Journal of the Air & Waste Management Association*, vol. 60, no. 5, pp. 568–573, May 2010, doi: 10.3155/1047-3289.60.5.568.

[6] O. US EPA, "Inventory of U.S. Greenhouse Gas Emissions and Sinks," Feb. 08, 2017. https://www.epa.gov/ghgemissions/inventory-us-greenhouse-gas-emissions-and-sinks (accessed Apr. 19, 2022).

[7] X. Tian, Y. Zhou, B. Morris, and F. You, "Sustainable design of Cornell University campus energy systems toward climate neutrality and 100% renewables," *Renewable and Sustainable Energy Reviews*, vol. 161, p. 112383, Jun. 2022, doi: 10.1016/j.rser.2022.112383.

[8] D. J. Olsen, N. Zhang, C. Kang, M. A. Ortega-Vazquez, and D. S. Kirschen, "Planning Low-Carbon Campus Energy Hubs," *IEEE Transactions on Power Systems*, vol. 34, no. 3, pp. 1895–1907, May 2019, doi: 10.1109/TPWRS.2018.2879792.

[9] N. Mirzaei Alavijeh, D. Steen, Z. Norwood, L. Anh Tuan, and C. Agathokleous, "Cost-Effectiveness of Carbon Emission Abatement Strategies for a Local Multi-Energy System—A Case Study of Chalmers University of Technology Campus," *Energies*, vol. 13, no. 7, Art. no. 7, Jan. 2020, doi: 10.3390/en13071626.

[10] Y.-F. Ho, C.-C. Chang, C.-C. Wei, and H.-L. Wang, "Multi-objective programming model for energy conservation and renewable energy structure of a low carbon campus," *Energy and Buildings*, vol. 80, pp. 461–468, Sep. 2014, doi: 10.1016/j.enbuild.2014.04.054.

[11] A. M. Sampat *et al.*, "Coordinated management of organic waste and derived products," *Computers & Chemical Engineering*, vol. 128, pp. 352–363, Sep. 2019, doi: 10.1016/j.compchemeng.2019.06.008.

[12] P. A. Tominac and V. M. Zavala, "Economic properties of multi-product supply chains," *Computers & Chemical Engineering*, vol. 145, p. 107157, Feb. 2021, doi: 10.1016/j.compchemeng.2020.107157.

292a Electrodialysis and Nitrate Reduction: An Electrochemical Reactive Separations Platform for Distributed Ammonia Recovery from Wastewaters

Jinyu Guo, Matthew J. Liu, Chloe M. Laguna, Sarah Blair, Brandon Clark, Dean Miller, Zach Perzan, Corisa Wong, Kuai Fang, Adam Nielander, David B. Lobell, Meagan S. Mauter, Kate Maher, Thomas F. Jaramillo, William Tarpeh

Motivation

Ammonia is an essential compound to modern day fertilizer production and chemical manufacturing. Global ammonia demand exceeds 150 million tons a year (market value 70 billion USD), with over 96% fulfilled by the Haber Bosch (HB) process. However, HB production plants are energy-intensive (1-2% annual global energy consumption), carbon-intensive (1.2-1.4% annual anthropogenic CO_2 emissions), and concentrated in developed countries, leading to inequitable pricing and distribution.[1-5] Electrifying ammonia production with renewable energy can help offset the energy and carbon-intensive HB process in a distributed fashion, where modular process units powered by renewable electricity leverage feedstocks to deliver ammonia at the source.

Wastewaters are an underutilized feedstock for electrified ammonia production. Large amounts of reactive nitrogen, such as HB ammonia, accumulate in the biosphere because 80% of wastewater globally is discharged without treatment.[6] Recovering nitrogen (N) from wastewater can simultaneously fulfill the roles of traditional removal technologies such as nitrification-denitrification and of fertilizer production processes such as HB. Two forms of reactive nitrogen dominate aqueous nitrogen emissions: ammonium (NH_4^+) and nitrate (NO_3^-). Wastewater ammonium can be recovered through processes that selectively separate it from other wastewater constituents. Meanwhile, wastewater nitrate can be recovered as ammonia through a selective electrocatalytic reduction reaction followed by a selective separation process. Ammonia recovery from ammonium- and nitrate-rich wastewaters therefore creates opportunities to co-locate selective reactions with selective separations. In this talk, we describe Electrodialysis and Nitrate Reduction (EDNR), a novel electrochemical reactive separation process that efficiently couples water treatment and ammonia production from wastewaters.

Methods

The EDNR reactor consists of three chambers and operates in two stages (**Figure 1a**). Wastewater influent enters the middle chamber while ammonia is electrochemically recovered in the left and right chambers. In Stage 1 (electrodialysis), influent and migrate to the left and right chambers, respectively, due to an applied potential. During this stage, hydrogen evolution in the right chamber creates an alkaline environment (pH ~ 12). As a result, influent ammonium deprotonates to yield ammonia (pKa ~ 9.25): $NH_4^+ \leftrightarrow NH_3 + H^+$. In Stage 2 (nitrate reduction), ammonia is selectively produced from the electrochemical nitrate reduction (NO_3RR) in the left chamber: $NO_3^- + 8e^- + 9H^+ \rightarrow NH_3 + 3H_2O$. Cycling between the two stages generates concentrated ammonia solutions in the right and left chambers. EDNR leverages electrochemical swings to facilitate engineering of the reaction environment in each chamber such that (1) process control is independent of the influent wastewater and (2) a broad portfolio of products can be recovered. For example, phosphorus (P) and/or potassium (K)-containing electrolytes can be used in the left and right chambers to allow production of different fertilizer blends by tuning the N-P-K content (the three macronutrients needed for plant growth).

Based on process and product needs, several materials were employed for specific purposes EDNR experiments. In the left chamber, Ti/IrO_2-Ta_2O_5 mesh was used for electrodialysis (Stage 1). Meanwhile, Ti foil was used as the NO_3RR electrode (Stage 2). Our group has previously characterized the near-surface structure of Ti foil under NO_3RR conditions and identified an optimal NH_3 Faradaic efficiency at

−0.8 V_{RHE}.[7] Ti/IrO_2-Ta_2O_5 mesh was used in the middle chamber as a counter electrode for NO_3RR. Lastly, Pt foil was used in the right chamber for electrodialysis. An anion exchange membrane separates the left and middle chambers, and a cation exchange membrane separates the middle and right chambers. Simulated wastewater influent (1.6 mM KNO_3 + 13.9 mM $(NH_4)_2SO_4$) and real wastewater (reverse osmosis brine) were used. Electrolytes in each chamber were recirculated in batch.

Aqueous speciation was quantified with cation chromatography, anion chromatography, and flow injection analysis (to measure total ammonia nitrogen: sum of aqueous ammonium and aqueous ammonia; indophenol method).

Results

During Stage 1 (electrodialysis), we demonstrate effective pH control in the left and right chambers (**Figure 1b**). The left chamber acidifies to a pH slightly above 1.5, which has been identified as optimal for NO_3RR on Ti.[8] The right chamber basifies to pH ~ 12, facilitating deprotonation of to yield ammonia. After three cycles of EDNR, >95% of influent is recovered as ammonia in the right chamber (**Figure 1c**). Meanwhile, ~100% of influent is electrochemically reduced to ammonia in the left chamber (**Figure 1d**). We found three factors that enabled full nitrate conversion to ammonia: (1) the choice of weakly adsorbing perchlorate in the supporting salt; (2) the use of a high concentration of sodium in the supporting salt; and (3) pulsed electrolysis to periodically refresh the interfacial electrolyte. Parametric analyses of additional operating parameters further optimize the energy efficiency, nitrate conversion, and ammonia recovery rate of EDNR.

References

(1) Smith, C.; Hill, A. K.; Torrente-Murciano, L. Current and Future Role of Haber–Bosch Ammonia in a Carbon-Free Energy Landscape. *Energy Environ. Sci.* **2020**, *13* (2), 331–344. https://doi.org/10.1039/C9EE02873K.

(2) MacFarlane, D. R.; Cherepanov, P. V.; Choi, J.; Suryanto, B. H. R.; Hodgetts, R. Y.; Bakker, J. M.; Ferrero Vallana, F. M.; Simonov, A. N. A Roadmap to the Ammonia Economy. *Joule* **2020**, *4* (6), 1186–1205. https://doi.org/10.1016/j.joule.2020.04.004.

(3) Kyriakou, V.; Garagounis, I.; Vourros, A.; Vasileiou, E.; Stoukides, M. An Electrochemical Haber-Bosch Process. *Joule* **2020**, *4* (1), 142–158. https://doi.org/10.1016/j.joule.2019.10.006.

(4) Fernandez, C. A.; Hatzell, M. C. Editors' Choice—Economic Considerations for Low-Temperature Electrochemical Ammonia Production: Achieving Haber-Bosch Parity. *J. Electrochem. Soc.* **2020**, *167* (14), 143504. https://doi.org/10.1149/1945-7111/abc35b.

(5) Comer, B. M.; Fuentes, P.; Dimkpa, C. O.; Liu, Y.-H.; Fernandez, C. A.; Arora, P.; Realff, M.; Singh, U.; Hatzell, M. C.; Medford, A. J. Prospects and Challenges for Solar Fertilizers. *Joule* **2019**, *3* (7), 1578–1605. https://doi.org/10.1016/j.joule.2019.05.001.

(6) Clark, B.; Tarpeh, W. A. Selective Recovery of Ammonia Nitrogen from Wastewaters with Transition Metal-Loaded Polymeric Cation Exchange Adsorbents. *Chemistry – A European Journal* **2020**, *26* (44), 10099–10112. https://doi.org/10.1002/chem.202002170.

(7) Liu, M. J.; Guo, J.; Hoffman, A. S.; Stenlid, J. H.; Tang, M. T.; Corson, E. R.; Stone, K. H.; Abild-Pedersen, F.; Bare, S. R.; Tarpeh, W. A. Catalytic Performance and Near-Surface X-Ray Characterization of Titanium Hydride Electrodes for the Electrochemical Nitrate Reduction Reaction. *J. Am. Chem. Soc.* **2022**, *144* (13), 5739–5744. https://doi.org/10.1021/jacs.2c01274.

(8) McEnaney, J. M.; Blair, S. J.; Nielander, A. C.; Schwalbe, J. A.; Koshy, D. M.; Cargnello, M.; Jaramillo, T. F. Electrolyte Engineering for Efficient Electrochemical Nitrate Reduction to Ammonia on a

Titanium Electrode. *ACS Sustainable Chem. Eng.* **2020**, *8* (7), 2672–2681. https://doi.org/10.1021/acssuschemeng.9b05983.

Figure 1 Caption. EDNR process and performance for 3-cycle experiment. Left and right chamber electrolytes: 1 M $NaClO_4$. Each cycle consisted of a 1-hr ED stage (constant current: 22.5 mA) and a 2-hr NR stage (pulsed potential, cathodic potential: –0.8 V_{RHE} for 10 s, and returned to OCV for 10 s). **(a)** EDNR process schematic. **(b)** pH vs. time in the left, middle, and right chamber electrode. **(c)** ED efficiency (mol of NH3 in right chamber / mol of ammonium initially present in middle chamber) vs. cycle number. **(d)** NR efficiency (mol of NH3 in left chamber / mol of nitrate initially present in middle chamber) vs. cycle number

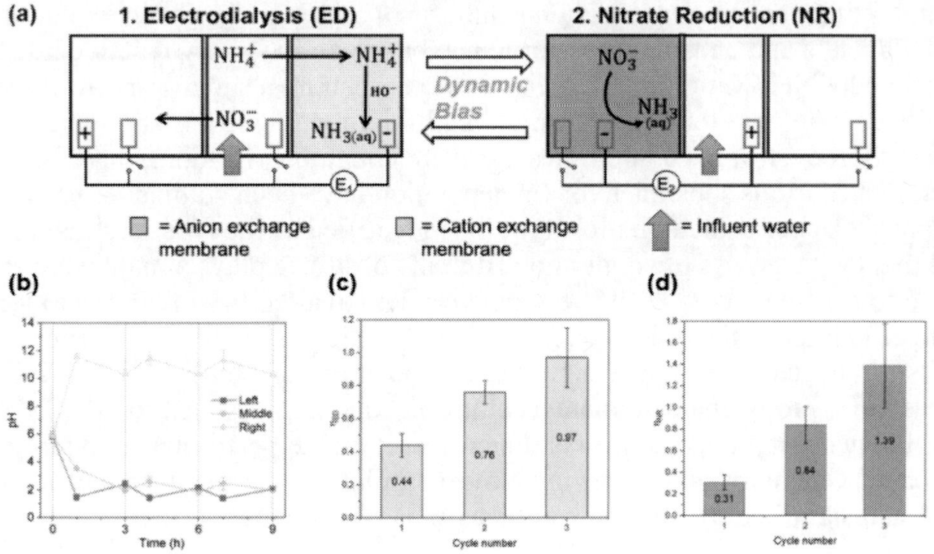

292b Calcium Mediated Electrochemical Synthesis of Ammonia

Ishita Goyal, Meenesh R Singh

The synthesis of green ammonia will be a steppingstone toward a sustainable and safe future. It not only finds applications as an agricultural fertilizer but is also being explored in energy storage domains. To sustainably utilize ammonia in these applications, alternate processes are required which can alleviate the massive carbon footprint and energy utilization, unlike the conventional Haber-Bosch process. Electrochemical synthesis of ammonia helps to make the process green by consuming renewable resources for electricity consumption. The proposed work aims to study electrochemical calcium-mediated ammonia synthesis (Ca-MAS). The reaction proceeds with the electrodeposition of calcium metal on the different substrates which subsequently undergoes nitridation and then finally leads to the formation of ammonia. Li-mediated synthesis of ammonia has been explored greatly by varying various parameters and enhancing performance but since Li is very reactive it is limited by safety issues associated with it. Calcium is expected to demonstrate a superior performance since it has a lower equilibrium potential for deposition, unlike lithium, and will assist in obtaining better energy efficiencies. Moreover, it is 10,000 times more abundant than Lithium, therefore, economically lucrative. Various substrates for Ca deposition have been studied to optimize the system's performance. Copper coated on nickel mesh is identified as the most promising substrate. It has been inferred that the thickness of copper on different substrates plays a major role and optimum thickness is needed to balance the trade-off between very less availability of copper to form the nitride and too much copper to impede the dinitrogen transportation. 50nm has been identified as the optimum thickness for this case. Various electrolyte compositions are studied with varying base electrolytes such as THF, propylene carbonate, acetonitrile, and Dimethyl sulfoxide (DMSO) based on their volatility, conductivity, and stability and their effect on the performance in terms of FE and current densities have been analyzed. Achieving almost similar yields and current efficiencies as Lithium-mediated will significantly lower the costs associated with the entire process and make it even more benign.

292c Green Ammonia-Mediated CO_2 Capture and Electrochemical Upgrading to Formate

Hengzhou Liu, Yifu Chen, Jungkuk Lee, Shuang Gu, Wenzhen Li

Direct electrochemical conversion of CO_2 capture solutions (instead of gaseous CO_2) into valuable chemicals can circumvent the energy-intensive CO_2 regeneration and pressurization steps. While commonly used CO_2 capture agents include alkali and amine solutions, ammonia has been rarely investigated. In another aspect, mismanagement of reactive nitrogen (Nr) in waste has emerged as a major problem in water pollution to our ecosystems, causing severe eutrophication and health concerns. Sustainably recovering Nr [such as nitrate (NO_3^-)-N] and converting it into green ammonia (NH_3) could mitigate the environmental impacts of Nr and reduce the NH_3 demand from the carbon-intensive Haber-Bosch process, as well as a possible CO_2 capture agent due to its alkaline nature. In this talk, we will present our rencet research on integration of electrodialysis and electrocatalysis for ammonia synthesis from dilute waste Nr sources, and green ammonia-mediated CO_2 capture (to ammonium bicarbonate, NH_4HCO_3) and subsequent reduction to ammonium formate (NH_4HCO_2) as a new approach to CO_2 capture and utilization (CCU). We have demonstrated a record-high NO_3^--to-NH_3 performance in a scalable, versatile, and cost-effective membrane-free alkaline electrolyzer (MFAEL): an unprecedented NH_3 partial current density of 4.22 ± 0.25 A cm^{-2} with a faradaic efficiency of $84.5 \pm 4.9\%$. We also discovered that an ammonium bicarbonate (NH_4HCO_3)-fed electrolyzer with an anion exchange membrane (AEM) outperforms the state-of-the-art $KHCO_3$ electrolyzer with a bipolar membrane (BPM) owing to its favorable thermal decomposition property, which allows for a 3-fold increase in the *in-situ* CO_2 concentration, a maximum 23% increase in formate faradaic efficiency, and a 35% reduction in cell voltage by substituting BPM with the AEM. Our integrated process by combining NH_4HCO_3 electrolysis with CO_2 capturing by on-site generated green ammonia from the electro-reduction of nitrate in MFAEL has shown a remarkable 99.8% utilization of CO_2 capturing agent. Such a multi-purpose process may offer a sustainable route for the simultaneous removal of N_r wastes and streamlined CO_2 capturing and upgrading to valuable chemicals.

292d Design of Ordered Mesoporous Oxides for Plasma-Assisted Catalysis of Ammonia Synthesis

Sonia Arumuganainar, Stephane Sartzetakis, Michele Sarazen, Bruce Koel

Energetic electrons in dielectric barrier discharges (DBDs) have been proposed to collide with and activate the N_2 triple bond,[1] enabling ambient NH_3 synthesis powered by renewable electricity but plagued by low energy yields, even with heterogenous catalysts. Porous oxides exhibit similar N_2 conversions with or without supported metal nanoparticles,[2] indicating significant support contributions that need to be quantified. We synthesized SBA-15[3] (a silica-based ordered mesoporous material (OMM) with tunable pore sizes) to probe porosity effects on N_2 conversion and plasma properties (i.e., electron densities). γ-alumina has not been similarly studied due to complex OMM syntheses and their lower surface areas, despite its higher dielectric constant (that improves DBD-catalyst synergy) and acid sites (that are catalytically active and adsorb product NH_3). We coated SBA-15 with different γ-alumina loadings[3] (5-15 wt. % Al) for the desired surface functionality on an identical framework, which all displayed complete coverage of the SBA-15 surface and ordered porosity, as evidenced from FTIR and N_2 physisorption, respectively. A sharp decrease in pore size and surface area indicated 10 wt. % Al is the threshold loading for a relatively thin coating. This composite, alongside SBA-15, was pressed and sieved into 1-2 mm pellets to promote beneficial packing in the DBD reactor, in which they exhibited similar N_2 conversions. However, shielding NH_3 from decomposition was only demonstrated for the composite, as evidenced from temperature-swing desorption after reaction, similar to effects observed on zeolites.[4] Both OMM pellets exhibited higher N_2 conversions relative to pelletized powders of non-ordered oxides, consistent with previous results for silica.[5] The composite also showed higher conversion than commonly used beads (silica and γ-alumina) despite the fact that interparticle voids between the smaller OMM pellets likely inhibit plasma generation (i.e., higher breakdown voltage) hence a partial discharge. This is in agreement with the lower electron densities and N_2 conversions measured on pelletized commercial samples relative to beads. These results inform the rational design of OMM porosity and functionality to optimize plasma properties, catalytic activity, NH_3 uptake, and therefore the energy yield of DBD-assisted catalysis.

References

(1) Mehta, P.; Barboun, P. M.; Engelmann, Y.; et al. *ACS Catal* **2020**, *10* (12), 6726–6734.

(2) Chen, Z.; Koel, B. E.; Sundaresan, S. *J Phys D Appl Phys* **2022**, *55* (5), 055202.

(3) Babaei, Z.; Najafi Chermahini, A.; Dinari, M. *Chem Eng J* **2018**, *352*, 45–52.

(4) Rouwenhorst, K. H. R.; Mani, S.; Lefferts, L. *ACS Sustain Chem Eng* **2022**, *10* (6), 1994–2000.

(5) Gorky, F.; Guthrie, S. R.; Smoljan, C. S.; et al. *J Phys D Appl Phys* **2021**, *54* (26), 264003.

292e Direct HCN Synthesis Via Plasma-Assisted Conversion of Methane and Nitrogen

Nefeli Kamarinopoulou, Dionisios Vlachos

Wasteful loss of methane (CH_4) is reported in remote locations, either in the form of CH_4 emissions or flare gas. Plasma-assisted hydrogen cyanide (HCN) synthesis from dilute CH_4 streams is a promising route to methane feedstock underutilization. The potential of dielectric barrier discharge (DBD) nitrogen (N_2)/CH_4 plasma for decentralized catalyst-free selective HCN production at ambient temperature conditions paired with renewable energy sources, renders it compelling compared to existing processes. In this work, we evaluate the suitability of N_2/CH_4 DBD plasma for selective HCN production, investigating the correlation between operating parameters, HCN and ethane (C_2H_6) yield and energy consumption (E_{HCN}). We demonstrate that HCN selectivity is favored by low initial CH_4 concentration. For further insight into parametric effects on radical induced chemistry, we develop a microkinetic model, establishing that a simplified reaction network with a significantly reduced number of active plasma species can reliably replicate yield trends. We then discuss the reactions and crucial intermediates governing HCN and C_2H_6 synthesis across experimental conditions. Finally, we compare this process to current industrial ones, regarding E_{HCN} and CO_2 equivalent emissions, showing that N_2/CH_4 DBD plasma can achieve minimal CO_2 emissions.

292f In-Situ and Operando Methods to Understand Electrified Processes for the Production of Fuels and Chemicals

Thomas Jaramillo

Electrification provides an intriguing pathway for the sustainable production of fuels and chemicals. Much remains to be understood about electrochemical pathways in order to advance technologies to the point of broad-scale commercialization. This paper will focus on the design, development, and implementation of in-situ and operando techniques that can provide new insights into electrochemical reactions of interest, including hydrogen production, the electrochemical conversion of CO_2 to carbon-based fuels, and N2 to ammonia. Techniques include those based on attenuated total reflectance surface-enhanced infra-red spectroscopy (ATR-SEIRAS), *on-stream* inductively coupled plasma – mass spectrometry (ICP-MS), and other forms of characterization involving X-rays and neutron scattering.

292g CO2 Electrolysis As Unprecedently Low-Cost and Scalable Method for Isotopes Production

Magda Barecka, Joel W. Ager III

The transition towards the decarbonized, sustainable chemical engineering industry is challenging due to the fundamental reliance of the chemical sector on petrochemical feedstocks. Production of critical chemical building blocks by electrochemical methods, powered with renewable sources of energy, is a rapidly evolving technology that could drastically reduce CO_2 emissions from the chemical sector. However, it has been extremely challenging to showcase a cost-competitive chemical production by means of CO2 electrolysis (CO2R). We present here an alternative strategy: initial use of CO2R to generate an extremely high-value product, namely carbon dioxide stream enriched in the rare isotope of carbon 13C. We have discovered that direct air capture (DAC) of carbon dioxide and electrochemical reduction of CO_2 can be modified so they will concurrently produce a carbon dioxide enriched in the stable ^{13}C isotope (patent WO2022271103A2). Our technology is simple to implement, requires minimal modification of existing reactors and does affect their function, and will easily scale to the current production capacity of ^{13}C isotopically labeled chemicals and beyond. The techno-economic analysis shows that CO2R-based separation of ^{13}C isotope could drastically decrease the cost of production of ^{13}C enriched streams (by a factor of 100). The excellent economic case of electrolysis-based isotopes separation will support investment into electrochemistry in general, which will lead to the improvement of the electrochemical reactors performance, decrease the investment cost, necessary to penetrate chemicals and fuels production markets. The benefits of such a synergistic development would span across several sectors and be an important stepping stone in the pursuit of Net Zero 2050 goal.

References:

[1] Barecka and Ager. *Towards an accelerated decarbonization of the chemical industry by electrolysis*. ***Energy Adv.***, 2023, **2**, 268-279

[2] Patent WO2022271103A2

292h Membrane-Free Electrochemical CO_2 Conversion Using Bicarbonate Solutions

Zhexi Lin, Nathaniel Blake, Reza Mirshekari, Oyinkansola Romiluyi, Xueqi Pang, Suryansh Kabra, Lucas Cohen, Daniel Esposito

Integrating CO_2 capture and utilization using electrochemical methods is a promising approach to addressing current energy and climate challenges.[1, 2] Conventional gas-fed membrane-based CO_2 electrolyzers have shown promise in this aspect, but several key challenges must be overcome for successful commercialization: i) large pH gradients that result in conversion of CO_2 into (bi)carbonate at the cathode and subsequent release of CO_2 at the anode,[3] ii) the degradation of polymeric membranes,[4] and iii) the decreased performance of membranes and electrodes caused by deposition of impurity species.[5]

In this work, we demonstrate a design of packed-bed membraneless electrolyzers (PBME) with a series of alternating porous electrodes to simultaneously address these three issues. By flowing a bicarbonate electrolyte resembling a carbon-capture solution through proton-generating anodes and CO_2 reduction cathodes, the pH of the electrolyte was balanced between electrodes while CO_2 was generated immediately upstream of the cathode, as evidenced by in-situ colorimetric imaging measurements.[6] This helps minimize the back-conversion from CO_2 to (bi)carbonate that results in low CO_2 conversion per pass. We show that the PBME design is scalable and that CO_2 utilization is expected to exceed 80% in an optimized multi-cell system. By adding chelating agents to scavenge metal impurities and ultra-thin metal oxide layers to encapsulate the electrocatalysts, impurity deposition was minimized and the electrode stability was significantly enhanced, as evidenced by electrochemical and product measurements, as well as electrode characterization. We also evaluated operations at elevated pressures, where the performance is expected to be improved due to the higher solubility of CO_2, hence resulting in a larger amount of CO_2 that can participate in the reaction.

This work demonstrates a simple, affordable, and scalable electrolyzer design that will enable the decarbonization of energy-intensive industrial processes through electrification in a sustainable energy future.

References

(1) Gutiérrez-Sánchez, O.; Bohlen, B.; Daems, N.; Bulut, M.; Pant, D.; Breugelmans, T. A State-of-the-Art Update on Integrated CO2 Capture and Electrochemical Conversion Systems. *ChemElectroChem* **2022**, *9* (5), e202101540, https://doi.org/10.1002/celc.202101540. DOI: https://doi.org/10.1002/celc.202101540 (acccessed 2023/04/01).

(2) Wang, G.; Chen, J.; Ding, Y.; Cai, P.; Yi, L.; Li, Y.; Tu, C.; Hou, Y.; Wen, Z.; Dai, L. Electrocatalysis for CO2 conversion: from fundamentals to value-added products. *Chemical Society Reviews* **2021**, *50* (8), 4993-5061, 10.1039/D0CS00071J. DOI: 10.1039/D0CS00071J.

(3) Rabinowitz, J. A.; Kanan, M. W. The future of low-temperature carbon dioxide electrolysis depends on solving one basic problem. *Nature Communications* **2020**, *11* (1), 5231. DOI: 10.1038/s41467-020-19135-8.

(4) Jiang, S.; Li, Y.; Ladewig, B. P. A review of reverse osmosis membrane fouling and control strategies. *Science of The Total Environment* **2017**, *595*, 567-583. DOI: https://doi.org/10.1016/j.scitotenv.2017.03.235.

(5) Hori, Y.; Konishi, H.; Futamura, T.; Murata, A.; Koga, O.; Sakurai, H.; Oguma, K. "Deactivation of copper electrode" in electrochemical reduction of CO2. *Electrochimica Acta* **2005**, *50* (27), 5354-5369. DOI: https://doi.org/10.1016/j.electacta.2005.03.015.

(6) Pang, X.; Verma, S.; Liu, C.; Esposito, D. V. Membrane-free electrochemical CO2 conversion using serially connected porous flow-through electrodes. *Joule* **2022**, *6* (12), 2745-2761. DOI: https://doi.org/10.1016/j.joule.2022.11.003.

344a Valorization of End-of-Life Tires Using Microwave Technology: 6PPD Extraction and Conversion

Sean Najmi, Montgomery Baker-Fales, Pooja Bhalode, Brandon Vance, Esun Selvam, Dionisios Vlachos

Every year, billions of tires are sent to scrap yards worldwide, where they are either landfilled, milled to smaller particles, or used as fuel for rotary kilns. The tire antioxidant and antiozonant, N-(1,3-dimethyl butyl)-N'-phenyl-1,4-benzenediamine (6PPD) was recently shown to cause acute mortality in aquatic life in the ng/L concentration range. Thus, current end-of-life (EOL) tire management strategies are dangerous for the environment, and different valorization strategies are needed. A critical step in the overall valorization strategy is the removal of 6PPD from EOL tires to ensure it does not leach into the environment. Additionally, legislation will soon outlaw the incorporation of 6PPD in new tires. Future valorization will be needed to be powered by renewable energy to meet decarbonization goals. Microwave (MW) assisted heating is an attractive technology for this application as the carbon black inherent to EOL tires is an excellent microwave energy absorber and can enable fast heating. We investigate solvent-based extraction of 6PPD from waste tires, its subsequent conversion into value-added products, and the pyrolysis of crumb rubber, all under MW-assisted heating.

The solvent screening was first carried out in batch mode under MW heating to determine which solvent is most efficient at extracting 6PPD from waste tires. The MW extraction was subsequently scaled up to a continuous flow system where high volumes of crumb rubber could be extracted. Temperature measurements were carried out down the length of the reactor and on the reactor wall at various power levels in a static bed and under gas and solvent flow. Temperature models were derived to relate power level to the temperature within the rubber phase.

After extraction, 6PPD and the extracted crumb rubber were subsequently upgraded. 6PPD was catalytically converted into cyclohexane. The condensed product mixture from the reactor effluent contained various aromatic and terpene products. The carbon black remaining in the solid phase was characterized. Our work presents strategies for the full valorization of the tires while providing mechanistic insights.

344b Efficient Modular Plasma Microreactor for Upcycling of Plastic Waste Derivatives

Darien Nguyen, Fabio Cameli, Panagiotis Dimitrakellis, Brandon Vance, Dionisios Vlachos

The pervasive and ubiquitous plastic pollution leads to the inherent need for an eco-friendly solution to mitigate this issue. Herein, the oxidative functionalization of plastic waste derivatives, particularly long aliphatic alkanes, using non-thermal, atmospheric plasma processing is presented as a green, catalyst-free step for plastic waste upcycling. A modular plasma microreactor utilizing Ar/O_2 gas mixtures was employed to functionalize n-dodecane. C_{12} secondary alcohols and ketones emerged as the main products with high production rates due to enhanced mass transfer via microreactor assembly. Higher conversions resulted in the formation of difunctional oxygenates, oligomerized oxygenates, and lighter hydrocarbon products stemming from carbon cleavage. The residence time and plasma-liquid interface were optimized by modifying the gas and liquid flow rates to achieve competitive energy yields. A mixture of liquid n-alkanes obtained through catalytic hydrogenolysis of low-density polyethylene was tested using the plasma microreactor to confirm process applicability. The results hold promise for upcycling plastic waste degradation products at ambient conditions using green, catalyst-free plasma discharges.

344c Decarbonizing Lime Kilns at Pulp Mills Via Oxy-Fuel Combustion/Electrification.

Edgar Carrejo, Ruochen Wu, Hasan Jameel, Fanxing Li, Sunkyu Park, Joe Sagues

The pulp and paper industry represents 2% of global industrial greenhouse gas emissions, of which the vast majority are biogenically derived. Thus, these biogenic carbon emissions represent an opportunity for carbon-negative pulp and paper processing if captured and sequestered. Despite this potential, the integration of these technologies has yet to be deeply explored in the pulp and paper industry. Recovery boilers, biomass and multi-fuel boilers, and lime kilns are the primary sources of carbon emissions in an integrated pulp mill. The most significant contributor is the recovery boiler, followed by multi-fuel boilers, with lime kilns contributing the least. However, lime kilns have an advantage from a carbon capture standpoint because they have the highest concentration of carbon dioxide (CO_2) emissions and are capable of handling even higher concentrations. Therefore, research on the feasibility of retrofitting or re-designing lime kilns to oxy-fuel combustion or electrification has gained significant attention over the past decade due to its generation of high-purity CO_2 (> 90%) and potential for low-cost CO_2 capture. In addition, multiple Direct Air Capture companies are developing electrified kilns for calcium carbonate ($CaCO_3$) calcination, and thus this research is directly applicable to such endeavors. The project involves assessing the feasibility of operating pulp mill lime kilns under high CO_2 conditions to elucidate potential barriers to using oxy-fuel or electrified lime kiln technologies. First, a kinetic assessment for the calcination of lime mud (mainly $CaCO_3$) from a commercial chemical pulp mill under high CO_2 concentrations was carried out. Both constant heating rate (FWO) and isothermal kinetic assessments were performed using TGA under idealistic conditions. Second, insight from the kinetic assessments was used to design experiments with a three-zone splittable, high-temperature, electrified rotary tube furnace under more realistic conditions. The large, rotary tube furnace was confined to operate within conditions acceptable to the pulping industry. A 1D numerical model was then developed in KilnSimu to simulate the calcination baseline (15% CO_2 concentration) and the high CO_2 scenario (90% CO_2 concentration) using a commercial lime kiln as a reference. To demonstrate the technology, three kilograms of lime mud were processed in the rotary kiln under a high CO_2 scenario (90% CO_2), reaching an extent of calcination greater than 95% with 70 minutes of retention time. In a lime kiln operation, producing good quality burned lime is essential. Burned lime (CaO) quality is assessed by its residual $CaCO_3$ content, availability, and reactivity. The amount of CaO in the burned lime available for slaking represents the lime availability. Moreover, lime reactivity refers to the speed at which the CaO can be hydrated in the slaker, forming calcium hydroxide ($Ca(OH)_2$), which reacts with the sodium carbonate present in the green liquor to regenerate the sodium hydroxide (NaOH) needed in the pulping process. The obtained burn lime under high CO_2 calcination was causticized with a 12.5% sodium carbonate solution for 2.5 hours at 90 °C, confirming the NaOH production. Then, the precipitated $CaCO_3$ was processed again in the rotary kiln under high CO_2 conditions to prove the reproducibility of the calcium cycle. The extent of calcination for the second round was similar to the first result, with a CaO content above 95%. This first-of-a-kind study demonstrates the technical feasibility of repetitively calcining pulp mill-derived lime mud effectively into CaO under conditions expected in commercial-scale, electrified kiln technologies.

344d Programmable, Electrified, and Far-from-Equilibrium Thermochemical Reactions

Qi Dong, Liangbing Hu

Conventional thermochemical processes such as natural gas (CH_4) conversion and ammonia (NH_3) synthesis are often powered by fossil-fuel-based continuous heating, which are extremely carbon-heavy with severe CO_2 emissions. In addition, as the near-equilibrium continuous heating lacks rapid, precise and time-resolved control over the temperature profile (*i.e.*, the reaction temperature and timescale), many reactions also suffer from poor selectivity, low yield, and/or poor catalyst stability. In this talk, I will present a novel far-from-equilibrium operating technique for thermochemical reactions (Nature 2022, 605, 470-476; Cover article) that employs programmable heating and quenching (PHQ) via electrification (*i.e.*, Joule heating), which precisely controls the transient heating time (*e.g.*, 20–110 ms) and provides rapid temperature quenching from up to 2400 K to near room temperature for gas phase reactants in milliseconds, resulting in improved selectivity, energy efficiency, and catalyst stability. Using CH_4 pyrolysis as a model reaction, our catalyst-free PHQ process enables high selectivity to value-added C_2 products (> 75% *vs.*< 35% by the conventional catalyst-free method, and *vs.* < 60% by most conventional methods with optimized catalysts) at a significantly lower energy cost (reduced by > 80%). Our PHQ process added a new dimension of tunability to thermochemical synthesis *via* the time-resolved temperature profile, which creates a unique opportunity for us to employ active learning to rapidly optimize any target product (*e.g.*, C_2H_4) with much less experimental effort compared to conventional trial and error approaches. Beyond the homogeneous and endothermic CH_4 pyrolysis, we further demonstrate the utility of PHQ in NH synthesis, which is heterogeneous and exothermic. Under far-from-equilibrium conditions by PHQ, we achieve a stable NH_3 synthesis rate of ~6000 μmol/g/h under ambient pressure for over 100 h with a non-optimized Fe catalyst, which is among the highest even when comparing to literature reports utilizing optimized catalysts. We envision the PHQ technique can potentially address some of the most pressing issues in the chemical industry by enabling process intensification and decarbonizing the chemical manufacturing process using renewable electricity.

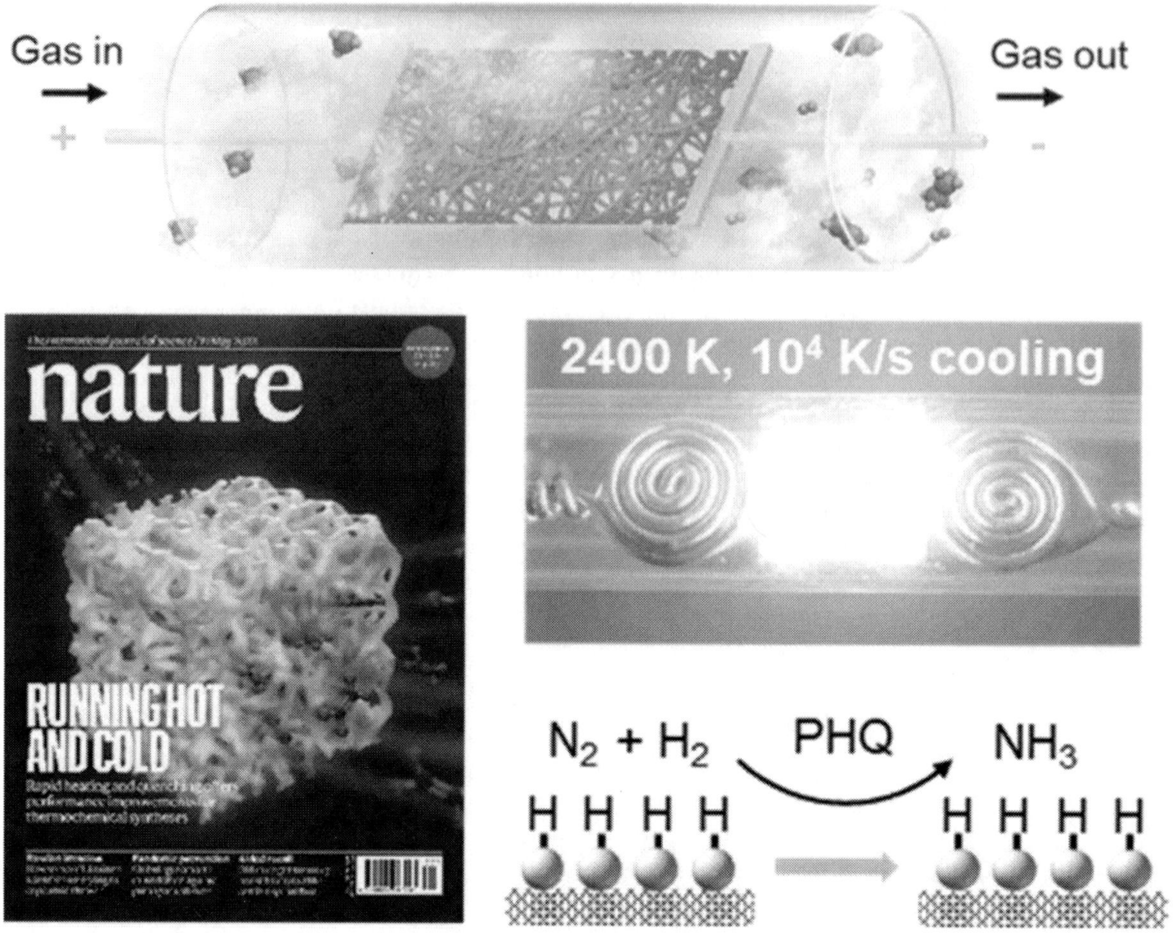

Gas in
+
Gas out
-
nature
RUNNING HOT AND COLD
2400 K, 10⁴ K/s cooling
N2 + H2
PHQ
NH3
H H H H
H H H H

344e Catalyst Deactivation Study during Microwave-Assisted Methane Dehydroaromatization

Victor Abdelsayed, Swarom Ravindra Kanitkar, Xinwei Bai, Daniel Haynes

Flaring of natural gas continues to be both an environmental and economic problem in the United States. Microwave-assisted reactions could offer a novel non-traditional approach in catalyst activation and product selectivity compared to a traditional thermal approach and can be developed into compact, modular systems for flare gas conversion at the well site. Due to thermodynamic limitation and poor catalyst stability for the methane dehydroaromatization reaction for aromatics production, understanding the catalyst deactivation would be beneficial towards designing a robust microwave catalyst. In this work, we studied the catalyst (Mo/HZSM-5) deactivation at different catalytic stages of performance from activation through deactivation. A comparison between microwave and conventional thermal reaction showed that the microwave-assisted methane dehydroaromatization reaction overperformed the conventional reaction towards aromatic yield but has shown to be less stable. These results could be due to formation of microwave hotspots on the catalyst surface inducing either more carbon formation or affecting the carbon due to the strong coupling which dehydrogenates the carbon to become more refractory and less reactive. In both cases this would cause rapid deactivation under microwaves. Different characterization tools were used to determine the activity-structure relationship, such as XPS, TPO, SEM-EDS, NH3-TPD, XRD, and dielectric properties for fresh and reacted catalysts.

344f Efficient Conversion of Ethanol to Acetaldehyde with Induction Heating

Han Wang, John O'Donnell-Sloan, John Pham, Hannah Schutte, Nicholas Wang, Carlos L. Pueyo, Jian-Ping Chen, Erdem Sasmaz

The utilization of renewable biomass as an alternative resource has received great attention to relieve the reliance on fossil fuels, which can lead to the emission of greenhouse gases causing global warming. Besides, process heating from the combustion of fossil fuels in many industrial applications is also one of the largest sources of greenhouse gas emissions. Ethanol is a sustainable resource from biomass fermentation, and its non-oxidative dehydrogenation is a vital source to obtain H2 and acetaldehyde, which is a key intermediate for producing acetic acid, ethyl acetate, n-butanol, and 1,3-butadiene. Supported Cu catalysts have been reported to show excellent ethanol conversion and selectivity to acetaldehyde at 280 °C. Since ethanol dehydrogenation to acetaldehyde is highly endothermic, a higher temperature is needed to increase the conversion efficiency. However, this can lead to uneven temperature distribution in the catalyst bed, decreasing the selectivity to acetaldehyde resulting from the undesired side products.

Such problems can be solved by applying induction heating (IH). IH is proven to be an alternative heating method for chemical reactions using a varying magnetic field and is considered a sustainable heating method due to the possible direct usage of clean energy, such as wind, solar, and hydropower. Although being explored in various applications, from cancer treatment in the medical field to metallurgic manufacturing on an industrial scale, IH hasn't been studied in the catalysis field until recently. In our work, we apply IH to ethanol-to-acetaldehyde reaction with physically mixed Co powder as the susceptor in the catalyst bed to provide heating directly to the catalyst. Our results show that ethanol conversion by IH is 1.7 times higher than that by conventional furnace heating at a temperature as low as 215 °C, suggesting the high production and energy efficiency of using IH. When the reaction temperature was increased to 245 °C with the furnace, the resulting conversion matched well with that obtained with IH at 215 °C, showing that IH significantly improves the energy efficiency by lowering the reaction temperature by 30 °C. The ethanol-to-acetaldehyde reaction by the IH and furnace heating showed comparable activation energies of 24.1 kJ/mol and 19.77 kJ/mol, respectively. In addition, reactions with IH and furnace heating show similar acetaldehyde selectivity of 80% from 215 °C - 245 °C, showing that the external magnetic field (MF) doesn't affect the active sites on the catalyst. To study and further improve the catalyst's activity with IH, we provided a model in COMSOL Multiphysics to simulate the temperature distribution and design the configuration of the catalyst bed. The temperature gradient along the reactor length was decreased to 1.5 K with the new design. In addition, the energy consumption was only 145 W compared to 435 W with the old design. The new configuration significantly increased the conversion, maintained the selectivity, and slowed down the deactivation by almost two times. The model provided a method to design the catalyst bed configuration and predict the energy consumption with desired product and energy efficiency.

IH significantly improves the efficiency of ethanol-to-acetaldehyde reaction, with the well-mixed Co powder providing the heat directly from inside the reactor, eliminating the heat transfer difficulties, quickly responding to the local temperature drop due to the highly endothermic reaction, and maintaining the stable reaction temperature. Therefore, it has great potential in highly endothermic and exothermic reactions, reducing carbon footprint, and further applications in producing value-added chemicals.

133a Multi-Period Heat Integration for Chemical Plant Electrification with Energy Storage

Mengdi Li, Faruque Hasan

To align with the net zero emissions goal, decarbonizing the chemical industry becomes crucial and inevitable. Industrial greenhouse gas (GHG) emission is the second largest source, taking 28% of the 2020 annual carbon emissions in the United States.[1] Electrification of the chemical industry enables the integration of cleaner energy sources such as renewables. However, the variation in renewable energy supply is a key challenge in electrifying industrial processes. Thermal energy storage (TES) utilizes low-cost storage mediums such as molten salt, rock, and sand that are non-toxic and more accessible, making it easier to scale up for industrial integration [2]. With the support from TES, renewable power electrical heaters can replace fossil fuel in chemical plants reducing at least one-third of the GHG emission in the manufacturing process.[2] Methods of mathematical optimization can be used to synthesize and design of utility systems. Papoulias and Grossmann (1983) proposed a MILP approach based on the LP transshipment model for the synthesis of flexible utility for changing process demand.[3] Multiperiod heat integration (MPHI) was developed to take into account process parameters varying over time. TES was introduced to MPHI problems to reduce energy demand and waste. Möhren et al. presented a simultaneous MILP model for TES integrating waste heat recovery [4]. Another TES-integrated MPHI was formulated into a non-convex MINLP model to optimize power cycle heat exchanger networks with different operation modes for renewables. [5]

So far, MPHI only considers several operating models for utility systems and energy storage. No research has time discretization involved in the multi-period model. There is a need for systematic methods that simultaneously consider the time-varying utility supply from renewables, the integration of TES with heat exchanger networks, and the scheduling of optimal charging/discharging operations. In this work, we first extend the LP transshipment model to minimize the TES size while maximizing the heat integration between hot and cold process streams under a time-varying renewable supply. Including year-wide energy scheduling with hourly resolution results in a large-scale optimization model. We propose a solution strategy to combine daily model with hourly resolution with yearly model with daily resolution to reduce the size of model.

The general problem for the multi-period heat integration with renewable utility and TES is constructed as the following. We are given a set of hot and cold stream inlet and outlet temperatures, flow rates, and heat capacity. Wind or solar power profile (Texas 2020 from ERCOT) is assigned to the plant in proportion to total generation with a constant price. Besides TES, the plant is equipped with a green hydrogen burner to encounter extreme situations. The objective is to minimize the levelized cost of the electrical heating (LCOEH) to obtain the optimal TES capacity, the amount of green hydrogen for backup heating, and hourly scheduling of TES charging and discharging. By varying the share proportion, we found that the variation in electricity supply dramatically influences the size of TES capacity and the use of backup green hydrogen. The LCOEH increased by 30% when the share of wind was reduced from 0.4% to 0.3%. The LCOEH is tripled in the solar case, further illustrating this conclusion. When the variation is too significant, burning green hydrogen during peak hours is more profitable than sizing up energy storage. Furthermore, we found that the increasing cost of green hydrogen affects the LCOEH to some extent depending on the variation in the electricity profile. The case study is extended to varying electricity prices by including real-time and day-ahead electricity price profiles in 8 locations in Texas. The flexibility in changing electricity demand allows the plant to take advantage of price variations, leading to lower energy costs. Future studies could consider participating in electricity market demand bidding to gain profit for flexible operations and alleviate increasing energy costs from electrification.

References

[1] EIA. (2020). Annual Energy Outlook 2020 with Projections to 2050.

[2] Zantye, M. S., Gandhi, A., Li, M., Arora, A., & Hasan, M. M. F. (2022). A systematic framework for the integration of carbon capture, renewables and energy storage systems for Sustainable Energy. Computer Aided Chemical Engineering, 2089–2094.

[3] EIA. (2021). Manufacturing Energy Consumption Survey 2018.

[4] Papoulias, S. A., & Grossmann, I. E. (1983). A structural optimization approach in process synthesis—I. Computers & Chemical Engineering, 7(6), 695–706.

[5] Möhren, S., Schäfer, C., Meyer, J., & Krause, H. (2022). A simultaneous approach for integration of thermal energy storages in industrial processes using multiperiod heat integration. Energy Storage and Saving, 1(2), 117–128.

[6] Elsido, C., Martelli, E., & Grossmann, I. E. (2021). Multiperiod optimization of heat exchanger networks with integrated thermodynamic cycles and thermal storages. Computers & Chemical Engineering, 149, 107293.

133b Emerging Electrosynthesis Processes to Decarbonize the Chemical Industry

Rong Xia, Ke Xie, Edward Sargent

Electrifying chemical productions is a potential approach to decarbonizing the chemical industry. Currently, 21% of the global greenhouse gas emissions comes from the industry sector, while chemical manufacturing accounts for the largest share of industrial carbon emissions. The electrochemical processes, when powered by renewable electricity, have lower carbon footprints than conventional thermochemical routes. The rapid development of renewable energy provides opportunities in electrifying chemical production and decarbonizing chemical industry. In this presentation, I would like to use electrosynthesis of value-added ethylamine and ethylene glycol as two examples.

First, I would like to talk about electroreduction of acetonitrile for selective synthesis of primary amines. Primary amines are important organic building blocks; however, the synthesis is often hindered by the poor selectivity because of the formation of secondary and tertiary amine byproducts. We report an electrocatalytic route to produce ethylamine selectively through an electroreduction of acetonitrile at ambient temperature and pressure. Among all the electrocatalysts, Cu nanoparticles exhibited the highest ethylamine Faradaic efficiency (FE, ~96%) at -0.29 V versus reversible hydrogen electrode (RHE). In a flow cell configuration, an ethylamine partial current density of 846 mA cm^{-2} was achieved using the Cu catalyst at -0.73 V vs. RHE in a 1 M NaOH electrolyte containing 12 wt.% acetonitrile. Moreover, the reaction mechanism of acetonitrile electroreduction was investigated by operando electrochemical differential mass spectroscopy and density functional theory calculations.

In the second part of this presentation, I would like to share our recent progress on ethylene glycol electrosynthesis. Ethylene glycol is mainly used as precursor of polyethylene terephthalate (PET) and antifreeze, with a world annual capacity of 42.09 Mt/yr. Conventionally, ethylene glycol is produced through ethylene partial oxidation at high temperature, followed by ethylene oxide hydrolysis, which has a significant carbon footprint. Here we report an electrochemical approach for ethylene glycol electrosynthesis with >90% selectivity towards ethylene glycol at a current density of 200 mA cm^{-2}. The reaction mechanism is investigated through operando differential electrochemical mass spectrometry experiment, isotopic-labeling experiment, and density functional theory calculation.

133c Bio-Oil Upgrading By Combining Chemical Pre-Reduction with Electrocatalytic Hydrogenation

Meheryar R. Kasad, James E. Jackson, Scott Calabrese Barton, Christopher M. Saffron

Fast pyrolysis of lignocellulosic biomass produces bio-oil, a mixture of several hundred compounds containing different oxygenated functionalities including alcohols, aldehydes, ketones, and carboxylic acids. Electrocatalytic hydrogenation (ECH) may be applied to produce fuels and chemicals from bio-oil under fairly mild conditions (ambient pressure and temperature < 80 °C). Promising results have been obtained for ECH of singular model compounds (e.g. furfural, 4-alkylguaiacols, di-p-tolyl ether) using noble metal electrocatalysts (ruthenium-platinum supported on an activated carbon cloth). However, ECH of multi-compound mixtures and bio-oil fractions remains a challenging prospect. It was observed during preliminary trials on whole bio-oil ECH that presence of compounds with an aldehyde functionality (e.g., furfural) inhibited the reduction of other compounds such as phenols and guaiacols. Therefore, ECH of a binary mixture containing a phenolic compound (4-propylphenol) and a heteroaromatic aldehyde (furfural) was investigated in the present study. Reagent-based reduction of the aldehyde group to an alcohol using sodium borohydride prior to ECH suppressed the inhibitory effect of the aldehyde group and improved reactivity of the phenolic compound during ECH. A similar approach involving chemical pre-reduction and ECH was subsequently extended to bio-oil fractions wherein it was observed that reduction of the aldehyde moieties to alcohols enabled conversion of the phenolic compounds present in the bio-oil fractions to cycloalkanols and cycloalkanes.

133d Electrification of Furanic Reduction Reactions: Insights into Balancing Desired and Undesired Reactions over Cu in Acid

Andrew May, Elizabeth Biddinger

Efforts to decarbonize the chemical industry have led to interest in renewable feedstock materials, such as biomass derived species, and a desire to couple reactions with intermittent renewable energy sources like solar or wind. Furfural is one biomass derived species that has been identified as a promising building block chemical due to its abundance and array of valorized products that can be formed. Furfural (FF) can be electrochemically reduced via hydrogenation to form furfuryl alcohol (FA), or hydrogenolysis to form 2-methylfuran (MF). FA is used to produce temperature stable resins and molds for foundry industry, and MF has been identified as a high octane fuel or fuel additive.

In acidic solutions on Cu electrodes, the electrochemical hydrogenation and hydrogenolysis reactions proceed in parallel reactions, showing that FA is not an intermediate in the formation of MF. For the selective production of the MF product on Cu electrodes, an acidic electrolyte is necessary, however it has been shown that FA and MF will undergo homogeneous side reactions in the acid leading to undesirable humin formation. At high concentrations of FF with long durations of electrolysis, fouling of Cu catalysts leading to performance drop has been shown as well. Understanding the different pathways to desired and undesired reactions that occur is key in designing catalysts and reactors that can increase the selectivity to desired products and minimize side reactions.

In this work, the characteristics and kinetics of desired and undesired reactions that occur during the electrochemical hydrogenation and hydrogenolysis of furfural on Cu in acidic solutions will be presented. We identified kinetic models for the desired electrochemical reactions towards FA and MF suggesting that they are formed via the non-competitive Langmuir-Hinshelwood mechanism through different intermediates. Undesired reactions occur both at the electrode, in terms of fouling, as well as homogeneously in electrolyte due to acidity. The undesired fouling reactions were found to be potential dependent, showing polymer formation or graphitic coking at moderate and high negative overpotentials, respectively. In solution, the undesired homogeneous reactions consumed FA and MF and were shown to follow first order kinetics with the proton activity and the concentration of FA or MF.

133e Achieving Order of Magnitude Increases in CO_2 Reduction Reaction Efficiency By Product Separations and Recycling

Akriti Sarswat, David Sholl, Ryan P. Lively

Decreases in renewable energy costs are driving significant interest towards electrochemical low-carbon fuel production strategies. Electrochemical conversion of CO_2 (CO_2RR) in e-refineries is at the center of many of these proposals. There have been numerous pieces of work focusing on reactor design and catalysts for these reactions, however, there are significantly less reports developing downstream separations processes and materials for the product mixtures which are inevitably needed to complement the reactions. The presented work focuses on designing downstream separations for such an e-refinery that converts CO_2 to ethylene with syn gas, acetic acid, and ethanol as by-products. We strategically utilize an input of methane to achieve >90% conversion of CO_2 directly captured from air or flue gas. The process has negative net carbon footprint and improves overall conversion efficiency of the process by an order of magnitude without any changes to the reactor itself.

133f Potential Benefits of Electrified Membrane-Distillation Hybrids over Individual Electrified Processes

Radhakrishna Tumbalam Gooty, Jose Adrian Chavez Velasco, Rakesh Agrawal

Molecular separations (separation of mixtures forming a *true solution*) are ubiquitous in chemical and petrochemical industries. Such separations are energy-intensive, so improving their efficiency is crucial for the reduction of the overall energy consumption of a chemical plant. Membrane-distillation hybrids are believed to reduce the energy requirement for separation dramatically. Therefore, in this work, we systematically investigate the potential that membrane-distillation hybrids hold in increasing the overall efficiency of separation of a binary non-azeotropic mixture. We begin with the description of a few physical insights that explain when and why is the hybrid more efficient than the individual separation technologies. Based on the insights, we then construct a superstructure that encompasses a wide variety of membrane-distillation hybrids. For this superstructure, we formulate a simplified optimization problem that identifies the most efficient hybrid arrangement. When desired, the solution to the optimization problem is further refined by carrying rigorous simulations in Aspen Plus. Using this framework, we analyze the potential of hybrids for three broad classes of separations: (1) close-boiling liquid separations, for which o-xylene/p-xylene is chosen as a representative mixture, (2) close-boiling gaseous separations, for which propylene/propane is chosen as a representative mixture, and (3) "easy" liquid separations, for which benzene/toluene is chosen as a representative mixture. Our analysis shows that the improvement in efficiency is highly subjective to the feed and product specifications. While the hybrids improved the efficiency substantially in some cases, there were also cases where the improvement in efficiency was very small to justify the increase in the complexity of the process. This implies that the hybrid arrangements are more efficient only in specific cases, and we note that the physical insights described here will help in identifying those cases.

133g Electrochemical and Chemical Mining of Municipal Solid Waste Incinerator Ash

Duhan Zhang, Michael Wang, Yet-Ming Chiang

Nowadays, in the U.S. more than 6.6 million tons of hazardous solid residue, called municipal solid waste incineration (MSWI) ash is annually generated in waste-to-energy (WTE) plants for producing electricity and reducing landfill volume. This ash cannot be further utilized for beneficial applications, so it is mostly buried and has a negative value with the cost of a disposal fee. Moreover, the economics of MSWI is challenged by the decreasing cost of renewable electricity. However, WSWI ash represents a unique opportunity for valuable element recovery, with the embodied value ranging from $100-400/tonne. We propose an electrochemical and chemical process for mining MSWI ash that utilizes electricity from the WTE plant to electrolytically produce acid and base streams for materials recovery and refining. We establish a sequential process of electrochemical extraction and chemical precipitation, and successfully recover Cu, Pb, Zn, Mg, Ca, Fe, Al, with a purified silica co-product. Recovery of >90% of the targeted elements at purities >90% is demonstrated. The industry-scale technoeconomic analysis shows our proposed technology will realize the possibility of extracting the embodied mineral value with net positive returns that significantly exceed the combined value of WTE electricity sales and landfilling cost.

31a Modeling and Techno-Economic Assessment of an Air-to-Syngas Process

Hussain Almajed, Omar J. Guerra, Wilson A. Smith, Bri-Mathias Hodge, Ana Somoza Tornos

Integrating carbon capture with utilization is one of the most promising pathways to pursue carbon neutrality[1–3]. Particularly, valorizing CO_2 by integrating Direct Air CO_2 Capture (DACC) with renewably-driven carbon electrolysis offers a promising pathway to overcome the high capture costs of DACC. More immediately, syngas production via integrated DACC-electrolysis routes can become economically feasible by 2050 as CO_2 electrolysis (CO_2ER) systems are well-suited to produce CO as a main product and H_2 as a byproduct. In this contribution, we study the integration of hydroxide-based DACC with a low-temperature CO_2ER and with conventional Reverse Water Gas Shift (RWGS) to identify improvement opportunities for DACC-Utilization (DACCU) in general and DACC-electrolysis in particular. We define the syngas H_2:CO molar ratio to be 2 and the production rate of syngas to be 2.2 Mt-syngas/yr, and we assume hydrogen supply from Alkaline Water Electrolysis (AWE) in both routes due to its higher technological maturity level. We present a verified Aspen Plus model of a hydroxide-based DACC plant and a custom-built model for CO_2 and water electrolysis to enable the assessment of the studied routes techno-economically. Herein, we refer to the integrated DACC with RWGS and with CO_2ER pathways as DACC-AWE-RWGS and DACC-AWE-CO_2ER, respectively.

Our findings indicate that DACC requires energy inputs of 8.58 and 8.55 GJ/t-CO_2 when integrated with RWGS and CO_2ER, respectively, in line with current literature estimates[4,5]. In addition, we calculate the power consumption of CO_2 electrolysis to be 6.56 MWh_{el}/kg-syngas and of H_2O electrolysis to be 6.01 and 9.68 MWh_{el}/kg-syngas for DACC-AWE-RWGS and DACC-AWE-CO_2ER, respectively. Such values, in addition to a referenced RWGS energy consumption value, are used to estimate the total energy consumption, and thus cost, of the two routes. Our assessment demonstrates that DACC-AWE-RWGS is a more carbon-efficient and a less energy-consuming pathway than the current state of the DACC-AWE-CO_2ER route. However, in terms of marginal energy-associated CO_2 emissions, we find DACC-AWE-CO_2ER to outperform DACC-AWE-RWGS, although at an insufficient emissions difference that would drive the shift from the conventional RWGS to the emerging CO_2ER. In addition, our CAPEX and OPEX calculations provide a lower total syngas cost of \$1.44/kg-syngas via DACC-AWE-RWGS compared to that of DACC-AWE-CO_2ER (\$1.53/kg-syngas). Further, our sensitivity analysis suggests that the H_2 production cost and the electricity price to be the main drivers in both pathways, and in that order. Moreover, our future scenario analysis compares three potential technology improvement scenarios and defines research targets for DACC-AWE-CO_2ER to compete economically with DACC-AWE-RWGS. Finally, we end with a discussion on the integration of renewable electricity with the investigated pathways.

References:

1. Mertens, J., Breyer, C., Arning, K., Bardow, A., Belmans, R., Dibenedetto, A., Erkman, S., Gripekoven, J., Léonard, G., Nizou, S., et al. (2023). Carbon capture and utilization: More than hiding CO2 for some time. Joule *0*. 10.1016/j.joule.2023.01.005.
2. IPCC (2022). Climate Change 2022: Mitigation of Climate Change: Summary for Policymakers (Cambridge University Press) 10.1017/9781009157926.001.

3. International Energy Agency (September 19). Putting CO_2 to Use (IEA Publications).
4. Keith, D.W., Holmes, G., St. Angelo, D., and Heidel, K. (2018). A Process for Capturing CO_2 from the Atmosphere. Joule *2*, 1573–1594. 10.1016/j.joule.2018.05.006.
5. Zeman, F. (2007). Energy and Material Balance of CO2 Capture from Ambient Air. Environ. Sci. Technol. *41*, 7558–7563. 10.1021/es070874m.

AUTHOR INDEX